Lecture Notes in Computer Science   11886

More information about this series at http://www.springer.com/series/7409

Uma Shanker Tiwary · Santanu Chaudhury (Eds.)

Intelligent Human Computer Interaction

11th International Conference, IHCI 2019
Allahabad, India, December 12–14, 2019
Proceedings

 Springer

Editors
Uma Shanker Tiwary 🆔
Indian Institute of Information Technology
Allahabad, India

Santanu Chaudhury 🆔
Indian Institute of Technology
Jodhpur, India

ISSN 0302-9743   ISSN 1611-3349 (electronic)
Lecture Notes in Computer Science
ISBN 978-3-030-44688-8   ISBN 978-3-030-44689-5   (eBook)
https://doi.org/10.1007/978-3-030-44689-5

LNCS Sublibrary: SL3 – Information Systems and Applications, incl. Internet/Web, and HCI

This Springer imprint is published by the registered company Springer Nature Switzerland AG
The registered company address is: Gewerbestrasse 11, 6330 Cham, Switzerland

Preface

The science and technology of Human Computer Interaction (HCI) has taken a leap forward in the past few years. This has given rise to two opposing trends. One divergent trend is to organize seperate conferences on focused topics, such as 'Autonomous Vehicles', 'Conversational Agents', etc., which earlier would have been covered under HCI. The other convergent trend is to assimilate new areas in HCI conferences, such as 'Computing with Words', 'Prosocial Agents Development', 'Attention based Applications', etc. The International Conference on Intelligent Human Computer Interaction (IHCI) is one of the rare conferences focusing on issues of 'Intelligence' and 'Human Computer Interaction' which exist at the cross-roads of the above mentioned two opposing trends. It is a privilege to present the proceedings of the 11th edition of this conference series (IHCI 2019), which was jointly organized by the Indian Institute of Information Technology in Allahabad (IIIT-A) and the Indian Institute of Technology in Allahabad, during December 12–14, 2019, at the IIIT-A campus.

Out of 73 submitted papers, 25 papers were accepted for oral presentation and publication by the Program Committee based on the recommendations of at least three expert reviewers. The proceedings are organized in five sections corresponding to each track of the conference as described below:

The 'EEG and Other Biological Signal Based Interactions' section is related to enabling technology that facilitates EEG signal analysis to understand and classify the brain states. In this track, five papers are presented incorporating concepts on EEG based motor imagery; post-stroke motor rehabilitation; classification of EEG signal; and biological signal treatment.

The 'Natural Language, Speech and Dialogue Processing' section is dedicated to exploring the area of dialog system; sentiment analysis in text and speech; and application of deep learning, transfer learning, and recurrent neural networks for classification of text and handwritten characters. Two papers on 'Abstractive Text Summarization' and a new approach called CWW (Computing With Words) are also included in this section. In total eight papers are presented in this section.

The 'Vision Based Interactions' section covers the research papers that focus on predicting body size using mirror selfies; Driver's Yawn detection; generating rules for cataract patient data; and retinal vessel classification. Four papers are presented in this section.

The 'Assistive Living and Rehabilitation and Interactions with Environment' section consists of three papers on topics as indicated in the title of the section itself.

Five papers are presented under the section 'Applications of HCI'. The papers cover topics on extracting community structure and virtual reality based training in a game scenario. The two apps presented in this section – one on applying ICT to intervene in Literary Disorders in the classroom and the other to support the science, technology, and innovation culture in youth groups – draw special attention.

We would like to thank the Organizing Committee for their immeasurable efforts to ensure the success of this conference. We would also like to express our gratitude to the Program Committee members for their timely and helpful reviews. Last but not least, we thank all the invited speakers, authors, and participants for their contribution in making IHCI 2019 a stimulating and productive conference. This IHCI conference series cannot achieve yearly milestones without their continued support in the future.

December 2019   Uma Shanker Tiwary

Organization

General Chairs

Santanu Chaudhury   IIT Jodhpur, India
Uma Shanker Tiwary   IIIT Allahabad, India

Program Co-chairs

Atanendu Sekhar Mandal   CSIR-CEERI, India
Debasis Samanta   IIT Kharagpur, India

Organizing Chair

Gaurav Harit   IIT Jodhpur, India

Steering Committee

Uma Shanker Tiwary   IIIT Allahabad, India
Tom D. Gedeon   The Australian National University, Australia
Debasis Samanta   IIT Kharagpur, India
Atanendu Sekhar Mandal   CSIR-CEERI, India
Tanveer Siddiqui   University of Allahabad, India
Jaroslav Pokorny   Charles University, Czech Republic
Sukhendu Das   IIT Madras, India
Samit Bhattacharya   IIT Guwahati, India

Program Committee

P. Nagabhushan   IIIT Allahabad, India
A. G. Ramakrishnan   Indian Institute of Science, India
Achard Catherine   Sorbonne University, France
Aditya Nigam   IIT Mandi, India
Akki B. Channappa   IIIT Allahabad, India
Alexander Gelbukh   Mexican Academy of Science, Mexico
Alexandre Vervisch-Picois   Télécom SudParis, France
Alok Kanti Deb   IIT Delhi, India
Amine Chellali   Evry Val d'Essonne University, France
Amrita Basu   JU, India
Anupam Agarwal   IIIT Allahabad, India
Anupam Basu   IIT Kharagpur, India
Ashutosh Mishra   IIIT Allahabad, India
Atanendu Sekhar Mandal   CSIR-CEERI, India

Bernadette Dorizzi	Télécom SudParis, France
Bibhas Ghoshal	IIIT Allahabad, India
Catherine Achard	Pierre et Marie Curie University, France
Daniel Wesierski	Gdańsk University of Technology, Poland
Daoudi Mohamed	IMT Lille Douai, France
David Antonio Gomez Jauregui	ESTIA, France
David Griol Barres	Carlos III University of Madrid, Spain
Debasis Samanta	IIT Kharagpur, India
Dijana Petrovska	Télécom SudParis, France
Ekram Khan	AMU, India
G. C. Nandi	IIIT Allahabad, India
Geehyuk Lee	KAIST, South Korea
Geoffrey Vaquette	Commissariat à l'énergie atomique et aux énergies alternatives, France
Gérard Chollet	CNRS, France
Jan Platoš	VŠB-TU Ostrava, Czech Republic
Jaroslav Pokorny	Charles University, Czech Republic
Jérôme Boudy	Télécom SudParis, France
José Marques Soares	Universidade Federal do Ceará, Brazil
K. P. Singh	IIIT Allahabad, India
Kavita Vemury	IIIT Hyderabad, India
Keith Cheverst	Lancaster University, UK
Laurence Devillers	LIMSI, Paris-Sud University, University of Paris-Saclay, France
Lopamudra Choudhury	Jadavpur University, India
Malik Mallem	Evry Val d'Essonne University, France
Manish Kumar	IIIT Allahabad, India
Maria Stylianou Korsnes	University of Oslo, Norway
Marion Morel	Pierre et Marie Curie University, France
Martin A. Giese	CIN/HIH University Clinic Tübingen, Germany
Michele Gouiffes	LIMSI, Paris-Sud University, University of Paris-Saclay, France
Mohamed Chetouani	Pierre et Marie Curie University, France
Mriganka Sur	MIT, USA
Nesma Houmani	Télécom SudParis, France
Partha Pratim Das	IIT KGP, India
Patrick Horain	Télécom SudParis, France
Plaban Kumar Bhowmick	IIT KGP, India
Pradipta Biswas	CPDM-IISC, India
Pritish K. Varadwaj	IIIT Allahabad, India
Harish Karnick	IIT Kanpur, India
Somnath Biswas	IIT Kanpur, India
Rahul Banerjee	BITS Pilani, India
Richard Chebeir	Pau et des Pays de l'Adour University, France
Samit Bhattacharya	IIT Guwahati, India

San Murugesan	BRITE Professional Services, Australia
Santanu Chaudhury	CSIR-CEERI, India
Satish K. Singh	IIIT Allahabad, India
Shekhar Verma	IIIT Allahabad, India
Shen Fang	Pierre et Marie Curie University, France
Sukhendu Das	IIT Madras, India
Suneel Yadav	IIIT Allahabad, India
Tanveer J. Siddiqui	University of Allahabad, India
Thierry Chaminade	Institut de Neurosciences de la Timone, France
Tom D. Gedeon	The Australian National University, Australia
Uma Shanker Tiwary	IIIT Allahabad, India
Vijay K. Chaurasiya	IIIT Allahabad, India
Vijayshri Tiwary	IIIT Allahabad, India

Local Organization Committee

Uma Shanker Tiwary	IIIT Allahabad, India
Shekhar Verma	IIIT Allahabad, India
Akki B. Channappa	IIIT Allahabad, India
Pritish K. Varadwaj	IIIT Allahabad, India
Vijayshri Tiwary	IIIT Allahabad, India
Ashutosh Mishra	IIIT Allahabad, India
Satish K. Singh	IIIT Allahabad, India
K. P. Singh	IIIT Allahabad, India
Vijay K. Chaurasiya	IIIT Allahabad, India
Bibhas Ghoshal	IIIT Allahabad, India
Manish Kumar	IIIT Allahabad, India
Suneel Yadav	IIIT Allahabad, India
Ajay Tiwary	IIIT Allahabad, India
Punit Singh	IIIT Allahabad, India
Rohit Mishra	IIIT Allahabad, India
Sudhakar Mishra	IIIT Allahabad, India
Shrikant Malviya	IIIT Allahabad, India
Santosh K. Baranwal	IIIT Allahabad, India
Pankaj Tyagi	IIIT Allahabad, India

Contents

Applications of HCI

EEG and Other Biological Signal Based Interactions

Classification of Motor Imagery EEG Signal for Navigation of Brain Controlled Drones

Somsukla Maiti[1,2(✉)], Anamika[1], Atanendu Sekhar Mandal[1,2], and Santanu Chaudhury[3]

[1] CSIR-Central Electronics Engineering Research Institute, Pilani, Rajasthan, India
somsuklamaiti@gmail.com
[2] Academy of Scientific and Innovative Research (AcSIR), Ghaziabad, India
[3] IIT Jodhpur, Karwar, Rajasthan, India

Abstract. Navigation of drones can be conceivably performed by operators by analyzing the brain signals of the person. EEG signal corresponding to the motor imaginations can be used for generation of control signals for drone. Different machine learning and deep learning approaches have been developed in the state of the art literature for the classification of motor imagery EEG signal. There is still a need for developing a suitable model that can classify the motor imagery signal fast and can generate a navigation command for drone in real-time. In this paper, we have reported the performance of convolutional stacked autoencoder and Convolutional Long short term memory models for classification of Motor imagery EEG signal. The developed models have been optimized using TensorRT that speeds up inference performance and the inference engine has been deployed on Jetson TX2 embedded platform. The performance of these models have been compared with different machine learning models.

Keywords: Motor imagery · Long short term memory · Convolutional stacked autoencoder · Drone · Jetson TX2

1 Introduction

With the emergence of drone technologies, drones have become an integral and significant component of almost every security and surveillance tasks in civil and military domain. In present scenario, the drones are controlled using handheld remote controller or joystick. This limits the usability of the drones for vast applications as the hands get occupied in the navigation operation. This in effect causes an increase in the effective workload of an operator that lead to under-performance or fatigue of the operator. Thus in order to design a suitable interface for controlling the navigation of a drone, a hands-free control is most desirable solution. In the past decade, researchers across the world have worked towards the development of different hands-free interfaces using human

© Springer Nature Switzerland AG 2020
U. S. Tiwary and S. Chaudhury (Eds.): IHCI 2019, LNCS 11886, pp. 3–12, 2020.
https://doi.org/10.1007/978-3-030-44689-5_1

interfaces such as body gesture, eye tracking and EEG signal. Brain controlled interfaces have gained significant attention due to the naturalistic way of human operation.

The need for development of a brain computer interface (BCI) for the navigation of drones has motivated the researchers to develop interface by analysing EEG signal corresponding to different facial expressions, such as, blinking, smiling, raising eyebrows etc., and mapping them to different motions of the drone. The EEG signal corresponding to these actions generate signal patterns that are significant enough to generate a command for the drone. However, the commands generated using facial expressions makes the interface very unreliable and unrealistic.

The most congruous solution to this problem can be achieved by analysing EEG signal corresponding to motor imaginations of the operator and generation of control signals for drone. Motor Imagery (MI) represents the mental process of simulating a motor action. These MI signals are primarily mediated by Mirror Neuron System which is primarily observed over the pre-motor cortex and sensory motor cortex. These signals share similar cognitive processes with actual motor movement execution. The MI signal is generally observed through the computation of suppression and enhancement of mu rhythms (8–12 Hz) while performing a motor imagery task [8].

Classification of motor imagery EEG signal has been performed using different machine learning and deep learning approaches since the last two decades. Owing to the fact that the EEG data is non-stationary in nature, most of the classification models are built with the consideration of adaptive parameter learning.

Most of the motor imagery classifications models developed uses a relevant set of features that can learn the necessary fluctuations in the EEG signal corresponding to the motor actions. To understand the underlying structure of the EEG signal, mostly Temporal features, such as, higher order statistical features (variance, skewness, kurtosis etc.), Morphological features, such as, Curve length, Average nonlinear energy, Number of peaks, Hjorth features (activity, mobility and complexity); Frequency and Time-Frequency Features, such as Fourier coefficients, Band Power, Wavelet Coefficients, Wavelet Entropy; Spatial Features, such as Common Spatial Patterns (CSP); and Connectivity Features, such as, Correlation or synchronization between signals from different sensors and/or frequency bands, Spectral Coherence etc. are computed.

Linear discriminant analysis (LDA), Support vector machines (SVM) and Bayesian classifiers have been most commonly used for classification of motor imagery signal for real-time operation. But to deal with EEG non-stationarity, adaptive classifiers has gained significant importance whose parameters are incrementally updated online. Adaptive LDA and Adaptive Bayesian classifier based on sequential Monte Carlo sampling have been developed for MI classification that explicitly model uncertainty in the observed data.

A significant number of researches have been performed using Riemannian geometry classifier (RGC) that maps the EEG data directly onto a geometric space or tangent space and learns a manifold [3]. The RGC computes intrinsic distance between data points in the manifold and classifies the data [9]. Although RGC can provide better generalization capabilities due to manifold learning, these class of classifiers are computationally extensive and thus is not suitable for a real-time online application.

2   Related Work

Dedicated feature extraction and selection requires a long calibration time and it changes for different sessions of data collection. Thus these models are unfit for developing a real-time BCI application. Deep learning methods are thus more suitable to learn the most relevant features for the task at hand. These models are also better suited to model high-dimensional data.

Convolutional Neural Network (CNN) based classifiers provide a better solution to this problem as it generates features of the signal by considering both temporal and spatial variation in the EEG data. Due to the intrinsic structure of using local receptive fields in convolution layers, the temporal variation in EEG automatically gets included. CNN based classifiers have been developed for the classification of motor imagery signal [7]. But due to the static structure of the convolution layers, it does not learn the non-stationarity in EEG data. Autoencoder (AE) based classifiers provide much better solution as it learns the features from the raw signal in unsupervised fashion and provide a good classification performance. Multi-modal stacked AE with separate AEs for EEG signals and other physiological signals has been reported to outperform the traditional classifiers. Sequential LSTM models and Bidirectional LSTM have been developed to learn the temporal variation in the EEG data.

Deep Belief Network (DBN) have been used extensively to classify motor imagery signals is another majorly used EEG classification method [2, 4–6]. DBN takes the non stationarity into consideration and learns features directly from data in hierarchical manner in unsupervised manner. Thus DBN classifiers outperform most of the other classification techniques. Sparse representation of the high dimensional EEG data can be obtained using sparse-DBN model. An adaptive DBN structure has been developed that can perform Belief Regeneration by using samples generated from a DBN to transfer the learned beliefs to a second DBN. In order to incorporate new knowledge in the model, the generated samples and new data from the stream can be used to train the DBN.

To develop a real-time BCI, we have proposed two deep learning models that learns the features of the EEG data through convolution and classifies the data in unsupervised fashion. The classifier models have been described in the following section.

3   Proposed Method

3.1   Data Description

We have used BCI competition IV data set 2b dataset for our experiments. The dataset consists of EEG data from 9 subjects performing two classes of motor imagery activities, viz., left hand and right hand movement. Five sessions of motor imagery experiments were performed on each of the subjects, among which the first two sessions were recorded without feedback and the other three sessions were recorded with online feedback. Each of the first two sessions contains 120 trials, and each of the rest three sessions (with feedback) includes 160 trials. The EEG data were recorded using three channels, C3, Cz, and C4 with a sampling frequency of 250 Hz.

3.2   Data Pre-processing

The EEG data have been first bandpass filtered between 0.5 Hz and 60 Hz, and a notch filter at 50 Hz was used to remove the power line noise. As the motor imagery occurs due to a change in mu-rhythm (8–13 Hz) and beta- rhythm (15–30 Hz) over the central motor cortex, the data is band pass filtered between 8–30 Hz. The data from the first 2 sessions are used for training the model.

Each experimental trial started with a fixation cross and a beep followed by a visual cue of 3 s provoking a left or right hand movement. The timing scheme of the motor imagery experimental paradigm has been shown in Fig. 1. The data between the period 4–7 s contain the motor imagery data for each trial, i.e., $(3 \times 250 =) 750$ samples of data. To incorporate the transition in data from EEG baseline to MI data, we have incorporated 100 ms of data at the start and end of the MI period. Thus the new motor imagery instances are $(0.1 + 3 + 0.1) = 3.2$ s long between the time period of 3.9 to 7.1 s for each instance, i.e., $(3.2 * 250) = 800$ samples of data.

Similarly, the non-MI data has been generated by considering data between duration 0–3.2 s from each experiment trial.

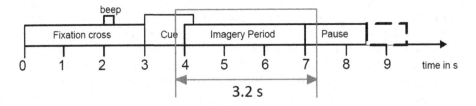

Fig. 1. Timing scheme of motor imagery

Thus we have considered three major classes for the classification problem, viz., Left MI, Right MI and Non MI. The EEG signals corresponding to left and right hand movement has distinct signature as shown in Fig. 2 which has been further classified.

All the studies on the dataset reports classification over Left and Right MI, but here we have considered Non Motor Imagery too as our purpose is to generate navigation commands for the drone corresponding to the thoughts of the operator.

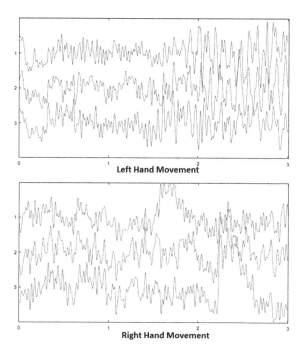

Fig. 2. Motor Imagery EEG for Left and Right hand movement

3.3 Data Augmentation

The number of left and right motor imagery and non-motor imagery data trials obtained from the database is very limited. The number of trials for each of the classes for the first two sessions are enlisted in the Table 1. Proper training of a deep learning architecture requires a voluminous dataset. With limited data availability, the learned model overfits and thus lacks the ability of generalization. Thus it is crucial to increase the volume of the dataset by augmentation of new data.

To increase the number of data instances, augmentation of EEG data has been performed. For generation of new realistic EEG data, we have a developed a generative adversarial network [1] architecture. The generator generates new EEG data from random noise using a convolutional network. The discriminator/critic, on the other hand, uses a convolutional architecture to discriminate the generated EEG data from the original EEG data. The similarity between generated and real data is computed using Wasserstein distance.

Wasserstein distance measures the distance between two probability distributions considering optimal transport cost. In Wasserstein GAN, the difference between the true distribution and the model distribution of the data is measured using Wasserstein distance instead of Kullback-Leibler divergence and Jensen-Shannon distance. This helps in alleviating vanishing gradient and the mode collapse issues in generator. WGAN continuously estimates the Wasserstein

Table 1. Number of trials per class for all subjects

	Left MI	Right MI	Non MI
Subject1	99	101	200
Subject2	99	99	198
Subject3	90	88	178
Subject4	123	123	246
Subject5	119	120	239
Subject6	77	84	161
Subject7	110	108	218
Subject8	121	101	222
Subject9	88	94	182

distance by training the discriminator to an optimal model. This helps in keeping the output in specific range even if the input is varied.

The structure of the generator and the discriminator has been shown in Figs. 3 and 4.

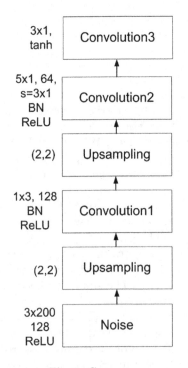

Fig. 3. Generator

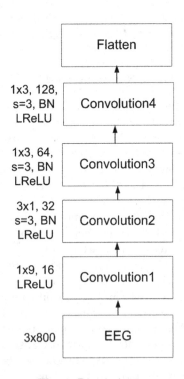

Fig. 4. Discriminator

3.4   Classification

Different dedicated feature extraction is generally performed to obtain signature from clean pre-processed EEG data which includes temporal, spectral, morphological features, viz., Hjorth parameters, Power Spectral Density (PSD), Curve length, number of peaks, Non Linear Energy, and Short Time Fourier Transform (STFT). To deal with the non-stationarity in the EEG data, STFT feature have been used. The STFT takes into consideration the temporal correlation in the data.

Instead of using a dedicated feature extraction module, we have used deep learning architecture to learn features from the EEG data in an automated manner (Fig. 5).

Fig. 5. Motor imagery classification with automated feature extraction

We have developed two models for classification of the motor imagery signal. To avoid the inter-person variability of EEG data, one classifier model was learnt for each of the subjects.

LSTM model learns the structure of EEG data by capturing temporal correlation in the data. As LSTM modules can choose to remember and discard the temporal information, it provides a better representation of the EEG data. Whereas the CNN SAE models learn a latent space representation of the data and does not specifically deal with the temporal variation of the data, which is the most prominent feature of EEG data.

Stacked autoencoder models map the data into a latent space and thus learns an unsupervised representation of the data. The convolution layers learn the local variation in the data. The features from the latent compressed representation are used for classification of the data.

3.4.1   Convolutional Stacked Autoencoder Model

A Convolutional Stacked Autoencoder (CNN-SAE) model, as shown in Fig. 6, has been developed to learn features from raw signal in unsupervised fashion and provide good classification performance. CNN-SAE combinatorial architectures provide the advantage of learning features using CNN and then learning the unsupervised SAE classifier on the detailed features.

Autoencoder learns a lower dimensional representation of the data and subsequent reconstruction of the data from the lower dimension space. The architecture of our network consist of two convolution layers followed by three autoencoder layers.

The EEG input is 3.2 s long, with the sampling frequency of 250 Hz that results in 800 samples for 3 channels. Therefore the input tensor is of dimension 800 × 3. The first convolution layer performs a temporal convolution over the batch normalized input tensor. 32 filters with kernel size of (1,30) are used. Resultant output is a tensor of shape (771,3,32). The second layer performs a spatial convolution with 64 filters having a kernel size of (3,1) in an overlapping manner giving an output of (78,1,64). Both convolution layers contain Parametric ReLU activation funcion (PReLU). The outcome after temporal and spatial convolution is flattened and fed to the first encoder layer containing 1200 hidden neurons. Two more successive encoders are employed with 600 and 150 hidden neurons. All the autoencoders have ReLU activation function and 25% of neurons are dropped after each encoder. Here the last encoder have the softmax activation function with three outputs indicating whether the input contains a Left MI, a Right MI or a non MI data.

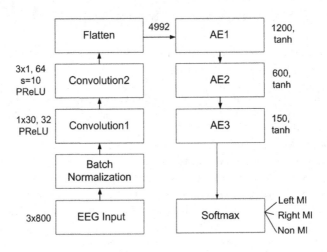

Fig. 6. Proposed CNN-SAE architecture

3.4.2   LSTM Based Model

To capture the sequential information in EEG data, LSTM based models have been developed for the classification. Three different models were developed for this purpose.

The first model is a shallow LSTM that uses two LSTM layers with 256 and 128 number of hidden layers respectively. The batch normalized input data is fed to the LSTM model where the output of each layer is passed through the tanh activation function. The output is passed through a dropout layer with dropout = 0.2 and recurrent dropout = 0.1 to reduce overfitting and improve the classification performance. The second model is a deep LSTM model with 256, 128 and 32 number of hidden layers. The same hyperparameter values are used for this model. The model was trained using RMSProp optimization while considering the categorical cross entropy loss.

The third model is a convolutional LSTM (ConvLSTM) model that learns the structure of the EEG data by capturing the spatio-temporal correlation in the data. The input to state and state to state transitions are computed using convolution. This model captures the variation in the data among different channels placed over the motor cortex as well as the temporal variation in the data.

4   Results and Discussion

The performance of different algorithms for Subject 4 have been reported in Table 2. The SVM and LDA are standard techniques used extensively for motor imagery classification. Extreme Learning Machine (ELM), regularized extreme learning machine (RELM) and Kernel-ELM (KELM) has been used to classify the motor imagery data due the fast computation of the algorithm which is very suitable for real-time EEG data classification. But the performance of unsupervised classification method DBN is the best as it learns the features from the data on its own. Adaptive generative models can be used for better online classification of the non-stationary EEG data.

Table 2. Motor imagery classification accuracy for Subject 4

Method	Accuracy (%)
SVM	93.008
ELM	92.461
RELM	87.961
KELM	91.980
LDA	91.880
DBN	93.751
CNN-SAE	**92.140**
LSTM	**95.330**
ConvLSTM	**92.530**

Although SVM provides a good classification performance, it does not provide a good solution for EEG data from another session. DBN model can classify the efficiently, similar to the CNN-SAE model. The deep learning models developed provide good accuracy at a higher processing speed. As the LSTM model considers the sequential variability in the data, it performs well for the motor imagery classification in real-time.

5   Inferencing on Jetson TX2 Platform

The developed algorithm has been deployed onto the embedded platform Jetson Tx2 which improves the performance and power efficiency using graph optimizations, kernel fusion, and half-precision FP16. The developed tensorflow model was optimized using TensorRT that speeds up inference performance. The inference time for each 800×3 block data is ≈ 9–10 ms. Thus the model has been further used to serially send navigation commands to the drone.

6 Conclusion

We have proposed a Conv-SAE model, a LSTM model and a Conv-LSTM model for the classification of the motor imagery EEG signal. Stacked autoencoder models map the data into a latent space and thus learns an unsupervised representation of the data. The convolution layers learn the local variation in the data. The features from the latent compressed representation are used for classification of the data. The LSTM models takes the temporal variation into consideration and thus both the spatial as well as temporal characteristics of the data are well considered. The developed LSTM model outperforms the CNN-SAE as well as the Conv-LSTM model. The developed models are well suited and generates navigation commands for drone after every 10 ms duration efficiently. We can further improve classification performance using ensemble models keeping the model complexity in consideration.

References

1. Arjovsky, M., Chintala, S., Bottou, L.: Wasserstein generative adversarial networks. In: International Conference on Machine Learning, pp. 214–223 (2017)
2. Bashivan, P., Rish, I., Yeasin, M., Codella, N.: Learning representations from EEG with deep recurrent-convolutional neural networks. arXiv preprint arXiv:1511.06448 (2015)
3. Congedo, M., Barachant, A., Bhatia, R.: Riemannian geometry for EEG-based brain-computer interfaces; a primer and a review. Brain-Comput. Interfaces 4(3), 155–174 (2017)
4. Lee, H., Grosse, R., Ranganath, R., Ng, A.Y.: Unsupervised learning of hierarchical representations with convolutional deep belief networks. Commun. ACM 54(10), 95–103 (2011)
5. Lu, N., Li, T., Ren, X., Miao, H.: A deep learning scheme for motor imagery classification based on restricted boltzmann machines. IEEE Trans. Neural Syst. Rehabil. Eng. 25(6), 566–576 (2016)
6. Movahedi, F., Coyle, J.L., Sejdić, E.: Deep belief networks for electroencephalography: a review of recent contributions and future outlooks. IEEE J. Biomed. Health Inform. 22(3), 642–652 (2017)
7. Tabar, Y.R., Halici, U.: A novel deep learning approach for classification of eeg motor imagery signals. J. Neural Eng. 14(1), 016003 (2016)
8. Wolpaw, J., Wolpaw, E.W.: Brain-Computer Interfaces: Principles and Practice. OUP, USA (2012)
9. Yger, F., Berar, M., Lotte, F.: Riemannian approaches in brain-computer interfaces: a review. IEEE Trans. Neural Syst. Rehabil. Eng. 25(10), 1753–1762 (2016)

Monitoring Post-stroke Motor Rehabilitation Using EEG Analysis

Shatakshi Singh[1](✉), Ashutosh Pradhan[2](✉), Koushik Bakshi[3], Bablu Tiwari[1],
Dimple Dawar[4], Mahesh Kate[4], Jeyaraj Pandian[4], C. S. Kumar[5],
and Manjunatha Mahadevappa[1]

[1] School of Medical Science and Technology, Indian Institute
of Technology Kharagpur, Kharagpur 721302, West Bengal, India
singh_shatakshi@iitkgp.ac.in
[2] Department of Electrical Engineering, Indian Institute of Technology Kharagpur,
Kharagpur 721302, West Bengal, India
ashutoshp012345@iitkgp.ac.in
[3] Advanced Technology Development Centre, Indian Institute
of Technology Kharagpur, Kharagpur 721302, West Bengal, India
[4] Department of Neurology, Christian Medical College, Ludhiana, India
[5] Mechanical Engineering Department, Indian Institute of Technology Kharagpur,
Kharagpur 721302, West Bengal, India

Abstract. EEG has proved to be a vital tool in diagnosing stroke. Now a days EEG is also used for tracking the rehabilitation process after stroke. Post stroke subjects are guided to do some physical activities to help them in regaining lost motor activity. The study presented here acquires EEG data from patients undergoing rehabilitation process and tracks the improvement in their motor ability. This is done by correlating change in Fugl-Meyer Assessment (FMA) score with some of the features obtained from EEG data. A significant correlation was found between FMA change and mean absolute value ($r = 0.6$, $p < 0.001$) of the EEG signal and also between change in mean alpha power ratio (left/right) vs. ΔFMA ($r = -0.71$, $p < 0.001$). The high correlation values of these features suggest that they can be used for monitoring rehabilitation after stroke.

Keywords: Stroke rehabilitation · EEG · FMA

1 Introduction

Stroke is a condition where blood supply to some part of the brain is obstructed. This leads to less oxygen and nutrients supply in brain tissues causing brain cells to die. Stroke can be ischemic or hemorrhagic. The death of brain cells in a particular part of the brain leads to loss of abilities controlled by that area of brain like loss of memory or loss of muscle control [1]. According to World Health Report, 2003 [2], stroke is one of the leading causes of disability in adults.

© Springer Nature Switzerland AG 2020
U. S. Tiwary and S. Chaudhury (Eds.): IHCI 2019, LNCS 11886, pp. 13–22, 2020.
https://doi.org/10.1007/978-3-030-44689-5_2

Stroke may lead to disabilities like limb motor function loss, speech impairment, etc. These can be partial or full depending upon the region of the brain it occurred. The loss of motor function is fatal as it affects activities of daily life (ADL). To improve the condition of these patients several therapies (occupational therapy, speech therapy, physical therapy) had been used in the past few decades. The idea behind these therapies is that repeated actions may lead to brain plasticity and hence to partial or full recovery of the patient [3].

A wide range of physical therapies are used that mimic daily life activities. Patients perform several exercises repeatedly with the help of physiotherapist [4]. These therapies can also be given with the help of some robot assisted automated system; where a system is trained to generate a particular action for specified time and the patient's limb moves along with the robot [5]. This type of system is useful as physiotherapist is not required and the programmed system will do all the work. Some systems even have virtual reality feedback that controls the motion of the robot according to the convenience of the patient. To make things more interesting some systems with interactive virtual reality(VR) games [6] and motion tracking device (MTD) are also developed. In Spite of all the efforts, we are still lacking in the field of efficiently tracking the progress of rehabilitation. According to a study it is seen that less than 15% subjects recover upper limb activity [7]. This limits both physical and social interaction of a patient. So, if we can come up with some efficient markers for tracking rehabilitation process then we can work towards our goal with a faster pace.

Electroencephalography (EEG)is one of the most powerful tools that can be used to diagnose cerebrovascular disease such as coma or seizures. EEG is useful for studying complex brain functions and has been known to provide insights of stroke effects [8]. EEG has proved to be a potential tool for tracking rehabilitation process. It is brain monitoring method that records the electrical activity of the brain. The patterns shown in EEG studies are related to electrical activity of the brain due to movements and different neuronal mechanisms related to motor control [9]. It is also shown that EEG data can be used to monitor brain plasticity due to various activities that a patient performs. So, analyzing brain signal of stroke patients by giving them some physiotherapy may be useful method for tracking rehabilitation [10].

The study presented here tracks the rehabilitation using three different domains. First, is the Power Spectral Density (PSD). In this Spectral density estimates are evaluated at frequencies which are logarithmic multiples of the minimum frequency resolution $1/T$, where T is the window length [11]. Second, some features obtained based on EEG data analysis for various domains. The features are obtained in time, frequency and time-frequency domains. Third, some Q-EEG parameters are calculated like delta to alpha ratio, brain symmetry index, etc. The analysis of the above parameters is done for evaluating rehabilitation progress.

2   Materials and Methods

2.1   Patients

Five post-stroke patients (male) were recruited for the study. All the patients had upper limb impairment in one hand only. The details about patients are listed in Tables 1 and 2. All the patients were informed and written consent was obtained. The study was thoroughly reviewed and approved by the Ethics Committee at Christian Medical College, Ludhiana. The experiment was conducted in the Department of Neurology, Christian Medical College, Ludhiana, India.

Table 1. Clinical data.

Variable	N = 5
Age ($mean \pm S.D.$) years	61 ± 10.6
Baseline Fugl Meyer Assessment ($mean \pm S.D.$)	94.8 ± 30.6
Diabetes	3(21.4%)
Hemoglobin, gm%	12 ± 2.3

Table 2. Patient information.

Patient number	Area of stroke
Patient 1	Frontal
Patient 2	Parietal/Occipital
Patient 3	Frontal/Parietal
Patient 4	Frontal
Patient 5	Central

2.2   Experimental Protocol

In this study patients were asked to sit in a relaxed position and a physiotherapist was present to help patients conduct the exercises properly. Patients were asked to perform a set of simple hand exercises with the help of physiotherapist. Each exercise was repeated 5 times with proper intervals in between. The details about exercises are given in Table 3. All the exercises were divided into two sections; first the subject has to do the exercise (Motor Execution ME) and in the next section was just imagining the action to be performed (Motor Imagery MI). Each session was about 2 h long. The exercises mentioned in Table 3 were repeated daily for 90 days and the EEG data was recorded at day 1 (baseline data), day 2, day 7, day 30 and day 90. Table 4 shows the protocol followed for each exercise.

Table 3. Exercise details.

1. Wrist extension −15°, Elbow 90°	12. Thumb opposition: MCP to PIP
2. Forearm neutral wrist flexion –hold for 5-s	13. Full finger abduction
3. Wrist extension and amp; Flexion (Slight)	14. Finger flexion- (a) Tip of Index finger to thumb tip
4. Wrist extension-15°, 0° - Elbow	(b) Tip of middle finger to thumb tip
5. Wrist full extension -Forearm pronation with 90° Elbow flexion	(c) Tip of ring finger to Thumb Tip
6. Forearm neutral -mass grasp (full extension to full flexion)	(d) Tip of little finger to Thumb Tip
7. Cylindrical grasp 5-Repetition	15. Mass grasp with squeeze ball
8. Grasp and release 5-Repetition	16. Holding pen with tripod grasp
9. Hook grasp	17. Wrist extension with resistance
10. Thumb abduction: 90°-0°	18. Finger abduction with resistance
11. Thumb flexion: 90°-towards palm	19. Holding paper b/w thumb and amp; fingers with resistance

Table 4. The protocol for each exercise.

Start	Trial 1	Trial 2	Trial 3	Trial 4	Trial 5	End	Rest (10 min)	Next Exercise

2.3   EEG Recordings

The EEG data was recorded pre and post exercise. Patients were asked to sit in a relaxed position with eyes closed for 2 min; eyes open for 20 s and 10 times eye blinking (Table 5). The signals were recorded using 64 gel based monopolar electrodes placed according to the 10/20 system. Impedance was less than 30 kΩ. The reference electrode was placed at Cz. The EEG data was acquired using G. Tec system with sampling frequency of 256 Hz.

Table 5. The protocol for recording EEG pre and post exercise.

Start	Eye close	Eye open	Eye close	Eye open	Eye close	Eye blink	Eye close	End
-	2 min	20 s	1 min	20 s	2 min	10 times	2 min	-

2.4   EEG Data Analysis

Data analysis was done offline using MATLAB R2015b and EEGLAB toolbox. While acquiring the data a band pass filter (0.5 Hz to 40 Hz, filter order = 8) and notch filter (50 Hz) were applied, and re-referenced using Common Average

Referencing (CAR). The reason behind using CAR is that when compared with other types of referencing methods, CAR reduces noise by greater than 30% [12]. Then baseline correction was done by removing the mean of recorded baseline data (1 s to 0 ms) from each channel. At last ocular artifacts (EOG) were removed using Independent Component Analysis (ICA) from EEGLAB tool box.

After preprocessing the data, analysis of was done in three main categories:

Power Spectral Density. The power spectral density (PSD) gives us an estimate of the frequency content of the signal. In this study we have compared PSD of each patient independently and monitored the response for all five days.

Feature Calculation (Time, Frequency and Time-Frequency Domain). The EEG signal is band passed into delta (1–4 Hz), theta (4–8 Hz), alpha (8–12 Hz) and beta (12–30 Hz) bands. The square of the sample amplitudes is summed to get the Energy in different bands and averaged over number of samples to get the Power. The Hjorth parameter Activity is determined as the variance of the signal and Mobility as the ratio of square root of variance of first derivate of the signal to variance of the signal. Activity is the total power of the signal and also the surface of power spectrum in frequency domain. Mobility is proportional to the standard deviation in the power spectrum and is an estimate of the mean frequency.

$$Activity = var(y(t)) \tag{1}$$

$$Mobility = \sqrt{\frac{var(dy(t)/dt)}{var(y(t))}} \tag{2}$$

where var () indicates variance and y(t) is the signal.

Table 6 lists all the features extracted in various domains.

Table 6. Features.

1. Energy $(\delta, \theta, \alpha, \beta)$ [13,15]	4. Mobility [13]
2. Mean absolute value [14]	5. Power beta [16]
3. Activity [13]	6. Power alpha [16]

Quantitative EEG (Q-EEG GUI). The Quantitative EEG GUI application is designed to compute the QEEG parameters such as mean delta, theta, alpha, and beta frequency powers, power ratio index, delta vs. alpha ratio and global brain symmetry index and compare the changes observed in the pre and post-baseline exercises [17]. The mean powers of each frequency band are displayed in tabular format comparing pre and post-exercise powers in the left and right hemispheres of the brain.

2.5   Statistical Analysis

Data distribution test was performed on the acquired raw EEG and found that the data is not normally distributed. Pearson's correlation coefficient between the calculated parameters and change in FMA (ΔFMA) was obtained with appropriate significance level. Also, comparison was done for the parameters before and after the exercise using Wilcoxon signed-rank test. The rank test was performed using IBM SPSS Statistics 20.0.

3   Results and Discussion

The PSD plots were analyzed for all patients in the post exercise base line activity with Eye closed (2 min). It can be clearly seen from the plots that as the

Table 7. (a)(b)(c) PSD and zoomed in PSD (alpha and delta) patient 2 O2 (d)(e)(f) PSD and zoomed in PSD (alpha and delta) patient 4 O2.

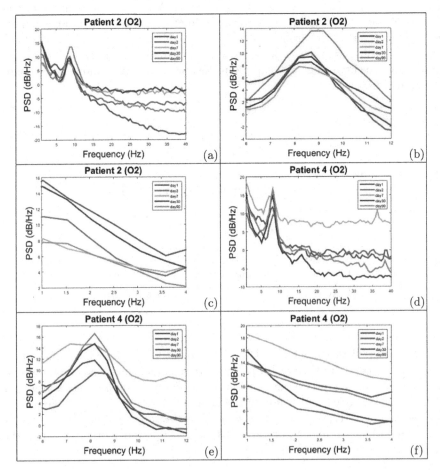

days' progress (day 1-blue dot to day 90-green star), there is a visible increase in alpha band power and decrease in the delta band power. The increasing power of alpha band and decreasing power in delta band marks the increased plasticity of brain cells due to physiotherapy as reported earlier in [19]. However, the degree of change was subjective to condition of each patient. Table 7 shows PSD of 2 patients (a, d) along with the magnified view of alpha and delta region (b, c, e, f).

Patient 2 had multiple lesions in the parietal, occipital and basal ganglia region. Table 8 shows the day-wise topo-plots of patient 2 from day 1 through day 90 at 8 Hz (lower alpha) from which we can clearly see the increase in activity in the lesion affected region as the days' progress.

The next part of the study correlates FMA score with the various EEG features calculated in different domains. A high correlation is seen between FMA change and mean absolute value of EEG signal (r = 0.6, p = 0.001) Table 9(a). As stroke patient's EEG signal is expected to have higher activity and lower mobility than a normal person, it can be seen that there is a negative correlation between change in FMA and change in the activity of EEG signal (r = −0.58, p = 0.002) Table 9(b) and a positive correlation is seen between Mobility of EEG signal change in FMA (r = 0.55, p < 0.01) Table 9(c).

Table 9(d, e, f) correlates some of the Q-EEG parameters with change in FMA. The change in mean alpha power ratio (left/right) vs ΔFMA shows statistically significant high correlation (r = −0.71, p < 0.001) Table 9(d). Similarly change in mean theta power ratio (left/right) vs ΔFMA shows a negative correlation −0.608 with a significance level of 0.00127 Table 9(f). This suggests increasing symmetry of the left and right hemispheres of the brain with respect to the alpha, beta and theta band power with the improvement in condition of the patient.

Table 8. Topo-plot of patient 2 showing day by day progress (in alpha band)

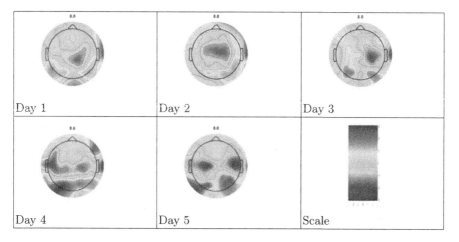

| Day 1 | Day 2 | Day 3 |
| Day 4 | Day 5 | Scale |

Table 9. Scatter plot showing correlation between change in FMA vs few parameters before exercise. (a) ΔActivity vs ΔFMA (b) ΔMean absolute value vs ΔFMA (c) ΔMobility vs ΔFMA.

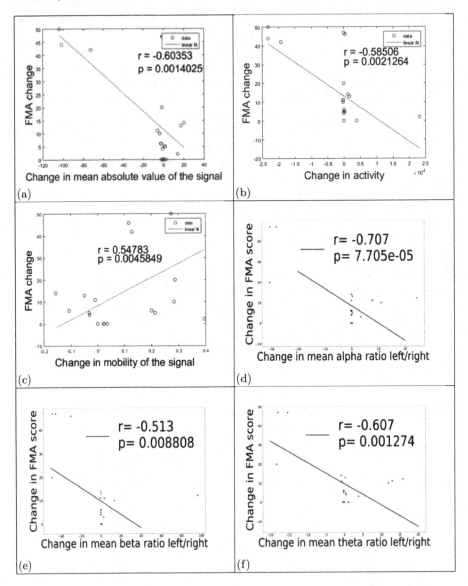

Two tailed Wilcoxon signed-rank test was also performed to check the hypothesis for changes in parameters before and after exercise. It was seen that for each parameter mentioned above; null hypothesis (H0 = H1) was rejected with 1% significance level. It shows that even just after the exercise significant changes were observed in each of the parameters considered in this study.

4   Conclusion

The work presented here extensively covered all the domains for EEG feature extraction. However, the features extracted are not exhaustive. Some of the previous reported features like revised Brain Symmetry Index, relative power, fuzzy approximate entropy show correlation in the range of 0.6 to 0.7 with a poor significance level [17,18]. The features used in this study showed strong correlation between the clinical FMA score and some of the calculated features with good significance level. A high correlation in the range of 0.81 to 0.99 was observed among the features too. So, these features can be used as agents for tracking rehabilitation progress in post stroke patients. It can also be established that if a patient continues to do exercises then it will be helpful in motor recovery. The study was done on a five subjects so a large scale study will be useful to further establish machine learning based methods to monitor rehabilitation progress.

Acknowledgements. This study is supported by the MHRD, Govt. of India, Dept. of Higher Education, Shastri Bhawan, New Delhi, and Indian Council of Medical Research, Dept. of Health Research, Ministry of Health and Family Welfare, Ansari Nagar, New Delhi - 110 029. (No: IIT/SRIC/SMST/RHU/2016-17/250 Dated: 10-04-2017)

Appendix

Institutional ethical clearance:
201808386/IECCMCL/RENEWAL-APPRVL/IMPRINT (Dated 28/8/2018).

References

1. About Stroke—American Stroke Association. https://www.stroke.org/en/about-stroke. Accessed 23 Oct 2019
2. The World health report: 2003: shaping the future. https://apps.who.int/iris/handle/10665/42789. Accessed 23 Oct 2019
3. Langhorne, P., Coupar, F., Pollock, A.: Motor recovery after stroke: a systematic review. Lancet Neurol. **8**(8), 741–754 (2009)
4. Luft, A.-R., et al.: Repetitive bilateral arm training and motor cortex activation in chronic stroke: a randomized controlled trial. Jama **292**(15), 1853–1861 (2004)
5. Colombo, R., et al.: Design strategies to improve patient motivation during robot-aided rehabilitation. J. Neuroeng. Rehabil. **4**(1), 3 (2007). https://doi.org/10.1186/1743-0003-4-3
6. Lin, B.-S., Chen, J., Hsu, H.: Novel upper-limb rehabilitation system based on attention technology for post-stroke patients: a preliminary study. IEEE Access **6**, 2720–2731 (2017)
7. Hendricks, H.-T., Zwarts, M.-J., Plat, E.-F., van Limbeek, J.: Systematic review for the early prediction of motor and functional outcome after stroke by using motor-evoked potentials. Arch. Phys. Med. Rehabil. **83**(9), 1303–1308 (2002)

8. Wu, J., Srinivasan, R., Burke, Q.-E., Solodkin, A., Small, S.-L., Cramer, S.-C.: Utility of EEG measures of brain function in patients with acute stroke. J. Neurophysiol. **115**(5), 2399–2405 (2016)

9. Platz, T., et al.: Multimodal EEG analysis in man suggests impairment-specific changes in movement-related electric brain activity after stroke. Brain **123**(12), 2475–2490 (2000)

10. Daly, J.-J., Wolpaw, J.-R.: Brain-computer interfaces in neurological rehabilitation. Lancet Neurol. **7**(11), 1032–1043 (2008)

11. Welford, B.-P.: Note on a method for calculating corrected sums of squares and products. Technometrics **4**(3), 419–420 (1962)

12. Ludwig, K.-A., Miriani, R.-M., Langhals, N.-B., Joseph, M.-D., Anderson, D.-J., Kipke, D.-R.: Using a common average reference to improve cortical neuron recordings from microelectrode arrays. J. Neurophysiol. **101**(3), 1679–1689 (2009)

13. Yucelbas, S., Ozsen, S., Yucelbas, C., Tezel, G., Kuccukturk, S., Yosunkaya, S.: Effect of EEG time domain features on the classification of sleep stages. Indian J. Sci. Technol. **9**(25), 1–8 (2016)

14. Harpale, V.-K., Bairagi, V.-K.: Time and frequency domain analysis of EEG signals for seizure detection: a review. In: 2016 International Conference on Microelectronics, Computing and Communications (MicroCom), pp. 1–6. IEEE(2016)

15. Geethanjali, P., Mohan, Y.-K., Sen, J.: Time domain feature extraction and classification of EEG data for brain computer interface. In: 2012 9th International Conference on Fuzzy Systems and Knowledge Discovery on Proceedings, pp. 1136–1139. IEEE (2012)

16. Al-Fahoum, A.-S., Al-Fraihat, A.-A.: Methods of EEG signal features extraction using linear analysis in frequency and time-frequency domains. ISRN Neurosci. **2014**, 1–7 (2014)

17. Mane, R., Chew, E., Phua, K.-S., Ang, K.-K., Vinod, A.-P., Guan, C.: Quantitative EEG as biomarkers for the monitoring of post-stroke motor recovery in BCI and tDCS rehabilitation. In: 2018 40th Annual International Conference of the IEEE Engineering in Medicine and Biology Society (EMBC), pp. 3610–3613. IEEE (2018)

18. Sun, R., Wong, W., Wang, J., Tong, R.-K.: Changes in electroencephalography complexity using a brain computer interface-motor observation training in chronic stroke patients: a fuzzy approximate entropy analysis. Front. Hum. Neurosci. **11**, 444 (2017)

19. Finnigan, S., van Putten, M.J.-A.-M.: EEG in ischaemic stroke: quantitative EEG can uniquely inform (sub-) acute prognoses and clinical management. Clin. Neurophysiol. **124**(1), 10–19 (2013)

Wavelet Transform Selection Method for Biological Signal Treatment

Gonzalo Jiménez[1][(✉)], Carlos Andrés Collazos Morales[1],
Emiro De-la-Hoz-Franco[2], Paola Ariza-Colpas[2],
Ramón Enrique Ramayo González[3], and Adriana Maldonado-Franco[4]

[1] Vicerrectoría de Investigaciones, Universidad Manuela Beltrán, Bogotá, Colombia
gonzalo.jimenez@docentes.umb.edu.co
[2] Departamento de Ciencias de la Computación y Electrónica,
Universidad de la Costa, Barranquilla, Colombia
[3] Laboratório de Sistemas Complexos e Universalidades, Departamento de Física,
Universidade Federal Rural de Pernambuco, Recife, Pernambuco, Brazil
[4] Universidad de la Sabana, Chía, Cundinamarca, Colombia
https://umb.edu.co/

Abstract. This paper presents the development and evaluation of an algorithm for compressing fetal electrocardiographic signals, taken superficially on the mother's abdomen. This method for acquiring ECG signals produces a great volumen of information that makes it difficult for the records to be stored and transmitted. The proposed algorithm aims for lossless compression of the signal by applying Wavelet Packet Transform to keep errors below the unit, with compression rates over 20:1 and with conserved energy in reconstruction as comparison parameter. For algorithm validation, the signal files provided by PhysioBank DataBase are used.

Keywords: Energy conservation · CR · ECG · Fetal · PDR · Wavelet packet transform

1 Introduction

The characterization of fetal electrocardiographic signals pretend to diagnose possible diseases that may put the life of the fetus at risk. To achieve these kind of characterization, technological tools have been proven useful [1]. Scientific literature shows there are invasive and non-invasive methods. The invasive method consists of placing an electrode on the scalp of the fetus, while the non invasive method consists of placing electrodes on the abdominal surface of the mother to capture the fetal ECG signal(electrocardiogram) [2]. In the latter case, the setup demands the detection and separation of signals from the mother and from the fetus. This is a difficult problem because these signals are surrounded by factors such as muscle movement noise, mother signal interference, electric noise produced by the measurement instruments and fetal movement. These factors change the signal power features, which is very low as it is [3].

© Springer Nature Switzerland AG 2020
U. S. Tiwary and S. Chaudhury (Eds.): IHCI 2019, LNCS 11886, pp. 23–34, 2020.
https://doi.org/10.1007/978-3-030-44689-5_3

The presence of noise in this signal and the interference of the signal from the mother make for a large volume of data and this issue should be considered in the signal characterization and analysis [4]. With a sampling frequency of 360 Hz, a record of 24 h requires 43 Mbytes. To achieve the desired compression level, according to the storage media needs and requirements or the transmission bandwidth, a threshold is determined for the type and amount of data. There are basically two compression methods: with loss or lossless [5]. As shown in most of the scientific literature, a compression with losses can achieve a compression rate above 20:1 but this leads to errors in reconstruction between 50% and 60%.

One of the challenges is selecting an adequate threshold to eliminate the noise levels in ECG signals [6], due to its non-linearity and this is particularly more important in fetal ECG signals. In this work, the amount of conserved energy in the reconstruction of the signal is used as a selection method for this threshold. This method allows improving the values of compression, keeping the errors under the percentage unit [7].

2   Theoretical Framework

The examination of the cardiac signals of the fetus is a tool used by obstetricians to monitor heart rhythm behavior, which is related to the well-being of the unborn [8]. These tests are performed from the 28th week of pregnancy. Among the methods for monitoring heart rate, we find the echocardiogram or DOPPLER but its use is questioned because this technique is too sensitive to fetus movements, which affects the obtained readings [9]. When using this method, there is also continuous signal loss [10]. Therefore, the use of electrodes on the abdomen of the mother is proposed for the detection of the fetal heart rhythm or FECG (Fetal Electrocardiogram)[11].

Figure 1 shows a block diagram of the signal treatment that is usually performed to obtain an ECG signal [12]. First, the signal is treated by a filtering, with the goal of eliminating electrical noise and separating the ECG signal from the mother. The main techniques for signal extraction are adaptive filtering and linear and non-linear decomposition [13,14]. Afterwards, the QRS complex is captured. Finally, the signal is processed, for example, applying wavelet [15,16]. Another technique is the blind separation of sources for the extraction of the features of the signals [17].

A comparative study of the compression methods shows that the search for simplicity in the coder allows lesser complexity in the compression, measured in the memory consumption for the process and the response time, but it is necessary to be careful when dealing with low-level signals that are embedded in noisy environments [18]. This is the case of fetal ECG signals, which are mixed with noise caused by the movement of the fetus, uterine contractions and that produced by the electrodes on the skin themselves. There are different types of thresholding that can be used in presence of non-white noise, but flexible thresholding and setting a threshold for each level present the least error rate [19]. Table 1 shows a comparison of the results of the compression rate (CR) and

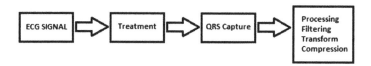

Fig. 1. Block diagram of ECG signal treatment

reconstruction error percentage (calculated by the Percent Root-Mean-Square Distortion or PRD) for different compression methods, signal types and used tools.

Table 1. Comparative results for different compression methods with the tools used in the process

Author	Loss	Fetal	Pre-processing	Thresholding	Quantization	Codification	CR%	PRD %
Jarisch (1980) [20]	With	Yes	Kalman filter	No	No	No	No	No
Mukhopadhyay (2013) [21]	With	No	T Hilbert	No	No	No	7.18	0.023
ZhiLin (2013) [2]	With	Yes	No	No	No	Bayesian learning	No	No
Jin Wang (2010) [22]	With	No	Wavelet	No	No	Neural network	7.6	2.74
Ebrajimzadeh (2011) [16]	With	No	Wavelet	No	Three levels of quantization	Huffman	13.92	0.97
Arvinti (2011) [23]	With	No	DWT Filter Bank	No	No	No	7.79	No
Honteng (2011) [24]	With	No	Wave atom	No	Two levels of quantization	Huffman	10.45	0.997
Jayashree (2011)	With	No	No	DWT	PDLZW	No	8.68	0.14
Hernando (2011) [25]	With	No	DWT	No	N-PR CMFB	No	0.58	No
Seong (2012) [26]	With	No	DWT	No	No	Compressed sensing	5.5	5.33
Zhicheng Li (2015) [27]	With	No	Wavelet	No	No	Compressed Sensing	4	
Chandan Kumar Jha (2016) [28]	With	No	Savitzky-Golay filter	No	Wavelet	Run Length Encoding	44	0.36
Chandan Kumar Jha (2015)	With	No	Band-pass and High-pass filter	No	Wavelet	x	90.23	0.34
Arvinti (2014) [4]	Without	No	No	No	Daubechies1 8	Daubechies1 8	14.98	No
Motinath (2016) [29]	Without	No	Saviky	No	Wavelet	WDT (db1-40)	15.2	0.23
Wang (2016) [30]	Without	No	No	No	Wavelet	Db1-16	14	0.97
Jiménez et al. (Proposed method)	Without	Yes	FIR filter	Universal	Wavelet	Db1-10	25	0.26

For several years, the Wavelet Transform has been used in different fields such as information compression and noise elimination [31]. The selection of the transform plays an important part in the analysis and processing of the signals [32]. To carry on with this task normally, a visual comparison is made between the signal to be treated and a series of transforms. Likewise, a correlation analysis can be performed between the signal and the transform. To be admissible, there are three requirements that every Wavelet Transform must meet. These are: a null average value, its energy must be finite and orthogonality. Mathematically, these conditions are expressed in Eqs. (1), (2) and (3).

$$\int_{-\infty}^{\infty} \Psi(t)dt = 0 \tag{1}$$

$$\int_{-\infty}^{\infty} |\Psi(t)|^2 \, dt < \infty \tag{2}$$

$$C_\Psi = \frac{1}{\sqrt{2\pi}} \int_{-\infty}^{\infty} \frac{|\hat{\Psi}(w)|^2}{|w|} dw < 0 \tag{3}$$

The main characteristics of a transform are: *The vanishing moments* of a Wavelet is related to the order of the function, so a function has N vanishing moments if it has an order of N. Mathematically, this is expressed as shown in Eq. (4).

$$\int_{-\infty}^{\infty} \Psi(t)^i dt = 0 \tag{4}$$

If (4) is true, then for $i = 0, 1, 2, ..., N - 1$, where i is the i-th vanishing moment. Also, the quantity of vanishing moments shows is related to the selectivity of the Wavelet decomposition. *The size of the support* is related to the order of the transform, therefore, with the vanishing moments like this: the size of the support is $2N - 1$. *The regularity* represents the softness of the wavelet transform. This helps for the reconstruction of the studied signal. *The scale function* is a function that is orthonormal to the wavelet function. If there is no scale function, it is not possible to use this wavelet family to discretize the signal.

Equation (5) shows a measurement parameter that considers the compression rate (CR). Basically, it consists of a comparison between the original signal and the compressed signal.

$$CR = \frac{L_{compressed\ signal}}{L_{original\ signal}} \tag{5}$$

And the similarity between the original signal and the reconstructed signal, given by Percent Root-Mean-Square Distortion (PRD), is shown in Eq. (6).

$$PRD(\%) = \sqrt{\frac{\sum_{n=1}^{N} (S_n - \tilde{S}_n)^2}{\sum_{n=1}^{N} S_n^2}} \tag{6}$$

where S and $\tilde{S}$ are the original signal and the reconstructed signal respectively.

The quality of the signal (QS), Eq. (7), is obtained from the relation between the compression rate and the reconstruction error percentage when comparing the original signal and the reconstructed signal.

$$QS = \frac{CR}{PRD} \tag{7}$$

3   Methodology

The ECG signals used in this study were obtained from PhysioBank DataBase [33] and the records last up to 4 h, sampled at 1 kHz by two channels. The files, other than the signal, also contain signal-related medical information. Each record consists of a header file, a signal file and a notation file. The available signal records consist of mixed maternal and fetal ECG signals, so they must be separated since the main input will be fetal ECG signals only.

Figure 2 presents a scheme of the steps of the proposed algorithm. The first step is filtering the signal to eliminate electrical noise and to establish a baseline. This filter is a FIR type of order 3.

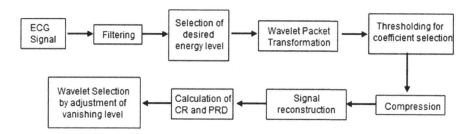

Fig. 2. Block diagram of the proposed algorithm

The method of blind source separation is applied to separate the ECG from the mother and the FECG. This method consists of separating two signals assuming they are mixed signals without knowing the separation distance between the sources and the sensors. It is like what happens when a certain sound can be distinguished and separated in an environment full of different sources without knowing the location of the source.

To compress the signal, it must be quantified. By using the discrete wavelet transform, R peaks of the fetal ECG signal are detected. The coding is done by a Huffman encoder. This type of encoder is chosen because this algorithm has a better performance at runtime although the compression ratio is not as high as in arithmetic coding. The selection of the transform within the Daubechies family is performed. The criteria is that the selected transform must allow the compression rate to reach the highest values. In this case, the goal is a value superior to 20:1. To achieve this result, the value for the retained energy in the reconstruction is set to 99.99%.

For the selection of the mother wavelet, the level of energy is set so it guarantees a PRD under 1%. The dispersion degree of the transforms applied to the registers' bandwidth is the criterion for the selection of the Mother Wavelet transform. This comparison considers the scale function and the amount of vanishing moments of the transform. The vanishing concept is related to how much the transformation resembles the original signal. In other words, this parameter indicates how well the applied transform maps the studied signal. The number of vanishing moments indicates the order of the transformation and the i-th moment is calculated as shown in Eq. (8):

$$\int_{-\infty}^{\infty} \Psi(t)t^n dt = 0 \quad with \; n = 0, 1, \ldots, N-1 \tag{8}$$

The highest possible order for the transform $\Psi(t)$ is N. The threshold allows selecting the coefficients that will be used for the compression. In this case, the used threshold is the universal threshold. In the search for a lossless compression, in this work, a hard threshold has been selected, as expressed in (9).

$$s(x) = \begin{cases} s(x) & |x| > \delta \\ 0 & |x| \leq \delta \end{cases} \tag{9}$$

where $s(x)$ is the function to be analized and δ is the proposed threshold. In the implementation of the thresholding method, a Perceptron type neural network is used as a classifier of the ECGF signal features. Once the thresholding method is selected, the neural network is implemented and trained to automatically select the threshold. By using a neural network, the signal energy thresholds will be set in order to discard the noise in the signal, trying to ensure that the information collected is reliable.

Then, signal compression is performed and the compression rate is calculated. The following steps are reconstructing the signal and calculating PRD. The measured parameter values are compared and, if necessary, the level of energy is adjusted to achieve a level of CR that is close but above 20:1, which is the goal of this work. As shown in Table 3, the transform family that provides the best performance for ECG signal compression is Daubechies [34]. Db2 results in CR lower than 10:1 and the maximum obtained value is 25:1 for Db6 and two decomposition levels. This order can vary from 1 to 45, but experimentally, it was proven that for order values greater than 6, the compression rate does not increase significantly, but PRD does begin to increase.

Having performed the analysis of the information, considering its relevance within the theoretical basis and evaluation of the results, the method is established. This implies the definition of the different conditions under which the derived method is valid. The validation is carried out by comparing the reconstructed signals with the original signals and to establish the degree of error in the reconstruction and the degree of compression of the signals. In this investigation, we propose a comparison that considers the amount of conserved energy after the reconstruction. This energy is affected by the levels of decomposition of the fetal ECG signals. This work uses as a transform selection base, the amount of energy conserved after the reconstruction of the signal. As observed in Fig. 3, a level of energy of 99.99% is maintained (Table 2).

Table 2. Properties of the Daubechies mother wavelet

Property	Daubechies
Order	1, 2, ..., 45
Regularity	Relative
Support size	2N − 1
Filter length	2N
Symmetry	Yes
Vanishing moments	N
Scale function	Yes
Continuous wavelet	Possible
Discrete wavelet	Possible
Explicit expression	No

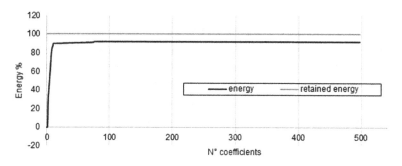

Fig. 3. Number of coefficients used related to the conserved energy in the reconstruction [11]

4   Results

Figure 4 shows that, as the signal decomposition level increases, CR increases as well. However, after order 6, the value is stabilized. With the decrease of the PRD, in comparison with Fig. 1, improvement is achieved in both performance parameters. In the case of CR, 25:1 was reached and PRD was decreased to 0.22%.

In Fig. 5, the comparison between the compression rate and the PRD for the obtained values in different records is displayed. This figure also shows how the compression rate values stay close to 20 in its majority, while the reconstruction error values are under the percentage unit. For some values, this is not the case

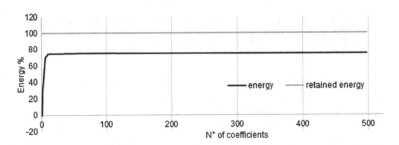

Fig. 4. Sample of CR compression value comparison, as a complement of the shown value of zeroes, with respect to PRD, identified as retained energy.

because they are the noisiest. This occurs because, as explained in the previous section, the wavelet selection requires keeping the most energy possible after the reconstruction. To evaluate this, wavelets of different orders were applied, from 1 to 10. Because of this, this order is chosen as the best. Figure 5 and Table 3 show that the proposed method offers better performance in both CR and PRD. The values of compression rate in Table 3 display the results of the parameters of compression performance, obtained by different authors along with those of the proposed method. This comparison considers the results and techniques applied to the same record. In the case of other records, the compression rate was affected by the noise level. However, in general, the signal compression is achieved under the conditions that were proposed at the beginning of the investigation and at levels that are superior to those of other authors. The quality factor (QS) is included even though it is not commonly used.

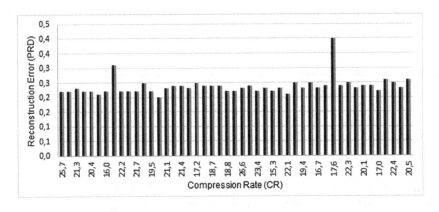

Fig. 5. Comparison between CR vs. PRD

Figure 6 shows the comparison between the original signal (red) and the compressed signal (black). The similarity of the compressed signal to the original can be appreciated.

Table 3. Comparative results of different compression methods

Author	Record	CR%	PRD%	QS
Lee (2011) [35]	100	9.6	0.44	21.8181818
Lee (2011) [35]	100	23	1.94	11.8556701
Ma (2015) [36]	100	15	0.29	51.7241379
Zhao (2016) [37]	100	14.8	7.58	1.9525066
Wang (2016) [30]	100	18.16	7.25	2.50482759
Jiménez et al. (proposed method)	100	25	0.22	113.636364

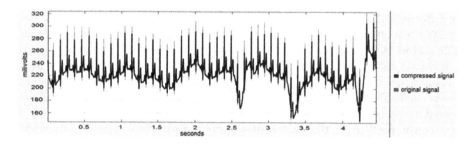

Fig. 6. Original signal vs. Reconstructed signal (Color figure online)

Finally, Fig. 7 shows a comparison of the results obtained by different authors on signal compression.

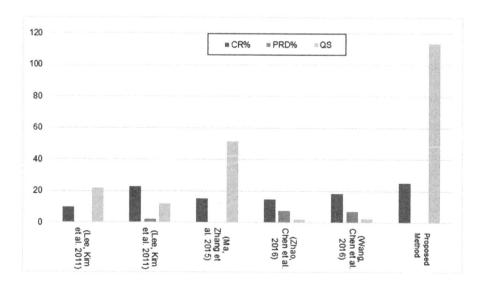

Fig. 7. CR, PRD and QS parameter comparison

5 Conclusions

To improve the compression rate of the signal, the method associated to Wavelet Transform that is mostly used, has been focused on the optimization on the selection of the decomposition tree [6]. In this work, the approach has been the selection of the best mother transform. The selection of the threshold of the conserved energy in the reconstruction presents a different method for the analysis of the signal to noise ratio, which is commonly used for the selection of the mother Wavelet applied to the compression.

As observed in Fig. 7, this method yields to improvement in the parameter levels when compared to others applied to the same record. The variation of the CR and PRD parameters depends on the required signal quality and the bandwidth or storage media that is available. The proposed method has a higher quality level (QS), because the PRD is under 1% and it is combined with a compression rate over 20, which reaches, in average, a value of 113.63 (as seen in column QS of Table 3). For storage, it is important that the compression rate is high. Meanwhile, for frequency analysis the compression rate can be lowered and keep a low PRD. The next step will be classifying through a neural network the coefficients that will adjust better to a certain type of pathology, observing its relationship with signal shape, in order to be able to increase the compression rate while keeping the original amount of data and the energy of the reconstructed energy as close to 100% as possible. The results show that the transform family that best performs in fetal ECG signal compression is Daubechies and the decomposition levels are between 2 and 4, to keep the CR and PRD values within the proposed ranges of 20:1 and 1% respectively.

References

1. Calabria, S.J.C., et al.: Software applications to health sector: a systematic review of literature. J. Eng. Appl. Sci. **13**, 3922–3926 (2018). https://doi.org/10.36478/jeasci.2018.3922.3926
2. ZhiLin, Z., Tzyy-Ping, J.: Compressed sensing for energy-efficient wireless tele-monitoring of noninvasive fetal ECG via block sparse bayesian learing. IEEE Trans. Biomed. Eng. **60**(2), 300–309 (2013). https://doi.org/10.1109/TBME.2012.2226175
3. Panifraphy, D., Rakshit, M., An efficient method for fetal ECG extraction from single channel abdominal ECG. In: IEEE International Conference on Industrial Instrumentation and Control (ICIC), Pune, India, pp. 1083–1088 (2015)
4. Arvinti, B., Costache, M.: The performance of the Daubechies mother wavelets on ECG Compression. In: 11th International Symposium on Electronics and Telecommunications (ISETC), Timisoara, Romania, pp. 1–4 (2014)
5. Brechet, L.: Compression of biomedical signals with mother wavelet optimization and best-biasis wavelet packet selection. IEEE Trans. Biomed. Eng. **54**(12), 2186–2192 (2007)
6. Castillo, E., Morales, D.P.: Efficient wavelet-based ECG processing for single-lead FHR extraction. Digit. Signal Process. **23**(6), 1897–1909 (2013)

7. Rivas, E., Burgos, J.C.: Condition assessment of power OLTC by vibration analysis using wavelet transformd. IEEE Trans. Power Delivery **24**(2), 687–694 (2009). https://doi.org/10.1109/TPWRD.2009.2014268
8. Bueno, M.C.: Electrocardiografía clínica deductiva, vol. 19. Universidad de Salamanca (2012)
9. Johnson, B., Bennett, A., Myungjae, K., Choi, A.: Automated evaluation of fetal cardiotocograms using neural network. In: 2012 IEEE International Conference on Systems, Man, and Cybernetics (SMC), Seoul, South Korea, pp. 408–413 (2012)
10. Ayres-de-Campos, D., Spong, C.Y., Chandraharan, E.: FIGO consensus guidelines on intrapartum fetal monitoring: cardiotocography. Int. J. Gynecol. Obstet. **131**, 13–24 (2015)
11. Kuzilek, J., Lhotska, L., Hanuliak, M.: Processing Holter ECG signal corrupted with noise: using ICA for QRS complex detection. In: 2010 3rd International Symposium on Applied Sciences in Biomedical and Communication Technologies (ISABEL), Rome, Italy, pp. 1–4 (2010)
12. Chen, Y., Cheng, C.: Reconstruction of sparse multiband wavelet signals from fourier measurements. In: International Conference on Sampling Theory and Applications (SampTA), Washington D.C., USA, pp. 78–81 (2015)
13. Rao, Y., Zeng, H.: Estimate MECG from abdominal ECG signals using extended Kalman RTS smoother. In: Sixth International Conference on Intelligent Control and Information Processing (ICICIP), Wuhan, China, pp. 73–77 (2015)
14. Yao, Z., Dong, Y.: Experimental evaluations of sequential adaptive processing for fetal electrocardiograms (ECGs). In: 49th Asilomar Conference on Signals, Systems and Computers, Pacific Grove, CA, USA, pp. 770-774 (2015)
15. Lima-Herrera, S.L., Alvarado-Serrano, C.: Fetal ECG extraction based on adaptive filters and wavelet transform: validation and application in fetal heart rate variability analysis. In: 13th International Conference on Electrical Engineering. Computing Science and Automatic Control (CCE), Ciudad de México, México, pp. 1–6 (2016)
16. Ebrahimzadeh, A., Azarbad, M.: ECG compression using wavelet transform and three-level quantization. In: 6th International Conference on Digital Content, Multimedia Technology and Its Applications (IDC), Seoul, South Korea, pp. 254-256 (2010)
17. Gerla, V., Paul, K.: Multivariate analysis of full-term neonatal polysomnographic data. IEEE Trans. Inf. Technol. Biomed. **13**(1), 104–110 (2009). https://doi.org/10.1109/TITB.2008.2007193
18. Chae, D.H., Alem, F.: Performance study of compressive sampling for ECG signal compression in noisy and varying sparsity acquisition, pp. 1306–1309. IEEE (2013). https://doi.org/10.1109/ICASSP.2013.6637862
19. Zhang, Y., Liu, B., Ji, X., Huang, D.: Classification of EEG signals based on autoregressive model and wavelet packet decomposition. Neural Process. Lett. **45**, 365–378 (2016). https://doi.org/10.1007/s11063-016-9530-1
20. Jarisch, W., Detwiler, J.S.: Statistical modeling of fetal heart rate variability. IEEE Trans. Biomed. Eng. (BME) **27**(10), 582–589 (1980). https://doi.org/10.1109/TBME.1980.326580
21. Mukhopadhyay, S.K., Mitra, M.: ECG signal processing: lossless compression, transmission via GSM network and feature extraction using Hilbert transform. In: Point-of-Care Healthcare Technologies (PHT), Bangalore, India, pp. 85–88 (2013)
22. Jin, W., Xiaomei, L.: ECG data compression research based on wavelet neural network. In: Computer, Mechatronics, Control and Electronic Engineering (CMCE), Changchun, China, pp. 361–363 (2010)

23. Arvinti, B., Isar, A.: An adaptive compression algorithm for ECG signals. In: IEEE 12th International Symposium on Computational Intelligence and Informatics (CINTI), Budapest, Hungary, pp. 91–95 (2011)

24. Hongteng, X., Guangtao, Z.: ECG data compression based on wave atom transform. In: IEEE 13th International Workshop on Multimedia Signal Processing (MMSP), Boston, MA, USA, pp. 1–5 (2011)

25. Hernando-Ramiro, C., Blanco-Velasco, M.: Efficient thresholding-based ECG compressors for high quality applications using cosine modulated filter banks. In: Engineering in Medicine and Biology Society (EMBC), pp. 7079–7082 (2011)

26. Seong-Beom, C., Young-Dong, L.: Implementation of novel ECG compression algorithm using template matching. In: 7th International Conference on Computing and Convergence Technology (ICCCT), Seoul, South Korea, pp. 305–308 (2012)

27. Li, Z., Deng, Y.: ECG signal compressed sensing using the wavelet tree model. In: 8th International Conference on Biomedical Engineering and Informatics (BMEI), Shenyang, China, pp. 194–199 (2015)

28. Jha, C.K., Kolekar, M.H.: Efficient ECG data compression and transmission algorithm for telemedicine. In: 8th International Conference on Communication Systems and Networks (COMSNETS), Bangalore, India, pp. 1–6 (2016). https://doi.org/10.1109/COMSNETS.2016.7439988

29. Motinath, V.A., Jha, C.K.: A novel ECG data compression algorithm using best mother wavelet selection. In: International Conference on Advances in Computing, Communications and Informatics (ICACCI), Jaipur, India, pp. 682–686 (2016)

30. Wang, X., Chen, Z.: ECG compression based on combining of EMD and wavelet transform. Electron. Lett. **52**(1), 1588–1590 (2016). https://doi.org/10.1049/el.2016.2174

31. Santamaría, F., Cortés, C.A., et al.: Uso de la Transformada de Ondeletas (Wavelet Transform) en la Reducción de Ruidos en las Señales de Campo Eléctrico producidas por Rayos. Información Tecnológica **23**, 65–78 (2012)

32. Craven, D., McGinley, B.: Energy-Efficient Compressed Sensing for Ambulatory ECG Monitoring, vol. 71, pp. 1–13. Elsevier, Amsterdam (2016). https://doi.org/10.1016/j.compbiomed.2016.01.013

33. Goldberger, A.L., et al.: PhysioBank, PhysioToolkit, and PhysioNet: components of a new research resource for complex physiologic signals. Circulation **101**(23), 215–220 (2000)

34. Singh, O., Sunkaria, R.K.: The utility of wavelet packet transform in QRS complex detection - a comparative study of different mother wavelets. In: 2nd International Conference on Computing for Sustainable Global Development (INDIACom), pp. 1242–1247 (2015)

35. Lee, S., Kim, J.: A real-time ECG data compression and transmission algorithm for an e-health device. IEEE Trans. Biomed. Eng. **58**(9), 2448–2455 (2011). https://doi.org/10.1109/TBME.2011.2156794

36. Ma, J., Zhang, T.: A novel ECG data compression method using adaptive fourier decomposition with security guarantee in e-health applications. IEEE J. Biomed. Health Inf. **19**(3), 986–994 (2015). https://doi.org/10.1109/JBHI.2014.2357841

37. Zhao, C., Chen, Z.: Electrocardiograph compression based on sifting process of empirical mode decomposition. Electron. Lett **52(3)**(9), 688–690 (2016)

Fuzzy Inference System for Classification of Electroencephalographic (EEG) Data

Shivangi Madhavi Harsha$^{(\boxtimes)}$ and Jayashri Vajpai

Department of Electrical Engineering, M.B.M Engineering College, J.N.V University, Jodhpur, India
shivangimadhaviharsha@gmail.com, jvajpai@gmail.com

Abstract. This paper aims to develop a Fuzzy Inference System that categorizes the Electroencephalographic (EEG) signals generated from a healthy brain with those generated by the brain suffering from epilepsy into different identifiable classes. This is done by statistical analysis of dynamical properties of the EEG signals using well established techniques for nonlinear time series analysis. For this purpose, a defined null hypothesis and obtained rejection counts are taken into consideration based on Randomness and Stationarity tests applied on the available EEG data. The outcome of this analysis is used to create a fuzzy inference model that can differentiate between brain states and zones they belong to viz. healthy zone, epileptic zone and non-epileptic zone. These zones are further classified into different states as eyes opened and closed for the healthy zone, the state where first ictal occurs, the seizure free interval and during the seizure activity within the epileptic zone and, the non focal and non epileptic state within the non–epileptic zone. Encouraging results have been obtained from this pilot project. However, limited data has been used for the development work and hence the decisions may be specific. Further generalization of the rules is possible with the help of more data and inputs from neurophysiological experts.

Keywords: Electroencephalographic (EEG) signals · Statistical analysis · Fuzzy logic · Fuzzy model · Epileptic EEG

1 Introduction

This paper brings together the knowledge of statistical analysis of dynamical properties and fuzzy inference systems for application to the field of classification of EEG signal in order to identify various states of the brain. The basic areas are now introduced.

1.1 Epilepsy and the EEG Signals

Electroencephalography (EEG) signals are one of the most widely analyzed and medically exploited signals that are generated from the human brain. Hence, examination of EEG signals to extract its dynamical properties plays an important role to identify with the brain's neuronal activities. It also helps to recognize and diagnose different neurological disorders. Epilepsy is one of the most widely occurring brain diseases that

© Springer Nature Switzerland AG 2020
U. S. Tiwary and S. Chaudhury (Eds.): IHCI 2019, LNCS 11886, pp. 35–48, 2020.
https://doi.org/10.1007/978-3-030-44689-5_4

affect human lives. Therefore, there is an ever-increasing need for development of new methodologies and technology that can help in management of this disease in a way that can make the lives of patients better and healthy.

Epilepsy is a condition consequential from underlying physiological abnormalities in which seizures represent an infrequent phenomenon to changing degrees. Clinically a seizure is an "intermittent paroxysmal, stereotyped disturbance of consciousness, behavior, emotion, motor function, perception, or sensation, which may occur singly or in combination and is thought to be the result of abnormal cortical neuronal discharges" [1].

The EEG is useful in the study of epilepsy because it can quantitatively show the changes of brain activity over time. As seizures represent a rare phenomenon, detecting these changes is important for the diagnosis and treatment of the disorder. The purpose of the detection of changes varies which may distinguish between seizure and inter-seizure EEG which is used for diagnosis.

1.2 Statistical Analysis of the EEG Signals

In order to conduct simple time series observations, this study aims to use formal statistical tests for nonlinear analysis using two tests namely Randomness test and Stationarity test. A suitable null hypothesis is formulated for the underlying EEG datasets for observation. The null hypothesis will be extensions of the statement that the data were generated by Gaussian linear stochastic processes. It is being carried out in attempt to reject a null hypothesis by comparing the value of a nonlinear parameter taken on by the data. This is estimated by using Monte Carlo resampling technique which is known as the method of surrogate data in the nonlinear time series literature [2].

The tests have been conducted by comparing the value of the relevant calculated parameter for the original signal with that of the surrogate data generated and based on the defined null hypothesis the test is said to be rejected or not rejected. A rejection at a given significance level means that if the null hypothesis is true, there is a certain small probability to still see the identified dynamical properties. Non-rejection means either the null hypothesis is true, or does not discriminate on the basis of calculated statistics. Less number of rejections is not able to provide distinguishable results for the comparison of data thus other speculations for analysis is sometimes required.

1.3 Fuzzy Inference System

Fuzzy inference system provides a method of mapping the given input variables to an output space via fuzzy logic based deduction mechanism comprising of Fuzzy If-Then rules, membership functions and fuzzy logical operations. Because the form of If-Then rule fits in human reasoning, and fuzzy logic approximates to people's linguistic habits, this inference process projecting crisp quantities onto human language and promptly yielding a precise value as result is gaining wide acceptance in intelligent decision making.

Over the past several decades, significant advances in neuro-imaging, genomic technologies, and molecular biology have improved the understanding of the pathological processes of epilepsy and the seizures occurring. Supplementary epilepsy syndromes have been delineated as a result of which existing International League Against Epilepsy

(ILAE) classification systems for seizures and epilepsies have become outdated and inadequate. This work thus contributes to extend the possibility of classifying different invasive and non invasive recordings of healthy persons and epileptic patients into different zones and brain states. This is done by proposing a method for implementation of fuzzy inference system. This autonomous model can be helpful for the neurologists to detect the brain states in a more convenient way.

2   Review of Literature

Significance tests have been developed which permit for the detection of nonlinearity to check results against the null hypothesis of a particular category of linear random processes [3]. This work has been studied in detail and followed in this research for comparison of extracranial (scalp) EEG signals of healthy volunteers and intracranial EEG recordings of epileptic patients.

One of the most accepted of such tests is the method of "surrogate data", studied for the nonlinear analysis by Schreiber and Schimtz [4] which can be used with any nonlinear datum that characterizes a statistic by one range. To test the null hypothesis that certain values of a nonlinear time series analysis measure calculated for the signals that are sufficiently explainable by properties compatible with a linear random process rather than by nonlinear deterministic dynamics, a collection of surrogate time series are generated having common linear properties, however are random. The null hypothesis is said to be rejected only if the result obtained for the original time series is outside the distribution of the surrogate collection at a given level of significance value.

Recently, Andrzejak *et al.* [5] showed that these findings carry over to bivariate signal analysis. Based on the same EEG recordings they showed that a combination of a nonlinear interdependence measure with bivariate surrogates excels nonlinear interdependence measures without surrogates as well as the linear cross correlation in localizing the epileptic focus. In both studies more rejections of the surrogate null hypothesis were obtained for the EEG recorded in the focal as compared to the non-focal EEG. The authors concluded that focal EEG signals were distinct from non-focal EEG signals in that they are less consistent with an underlying linear stochastic process and rather reflect some properties of a coupled nonlinear deterministic dynamics. However, as indicated above, the rejection of a surrogates' null hypothesis always leaves one with different alternative interpretations.

3   Methodology

This paper proposes to develop a fuzzy inference system that classifies different brain zones into identifiable categories based on statistical analysis of dynamic properties of EEG signals. The process of implementation includes selection of datasets, statistical analysis of the EEG signals and designing and implementation of the fuzzy inference system. These are explained in this section.

3.1 Dataset Description

The development of the Fuzzy Inference System was based on the statistical analysis of dynamical properties of seven datasets which have been taken from two different preprocessed and randomized EEG data sources. These have been widely referred in the published work of other researches [3, 4, 6] and are available in the public domain on the internet [7, 8] for facilitating research in this developing area. Each dataset analyzed for the dynamical properties comprises of 100 sample records of EEG recordings containing the original signal and for which surrogate data is generated for the randomness test and stationarity test. The number of times the conducted test was rejected against the defined null hypothesis was defined as the rejection count.

Seven datasets have been named as used in the resultant fuzzy inference model. Recordings from healthy volunteers in awake state with eyes opened are named HEO and that with eyes closed are named HEC. Recordings from non-epileptic zone from hippocampus formation in the opposite hemisphere are named NEZ. Recordings from non-epileptic zone where seizure focus is not found is labeled as NF. There are three datasets within epileptic zone. SFI are recordings during seizure free interval. EZF are recordings where ictals are first observed and are also called focal signals and recordings during seizure activity are named SZR. Exemplary EEG recordings for each dataset have been shown as follows (Figs. 1, 2, 3, 4, 5, 6 and 7):

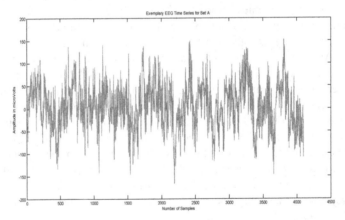

Fig. 1. Exemplary EEG time series from the recorded data for set HEO.

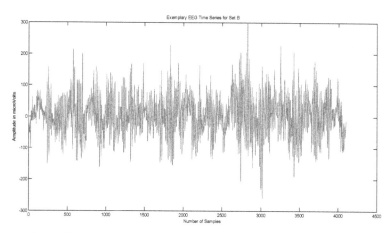

Fig. 2. Exemplary EEG time series from the recorded data for set HEC.

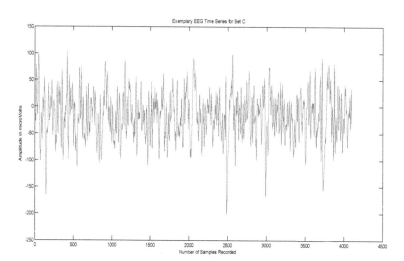

Fig. 3. Exemplary EEG time series from the recorded data for set NEZ.

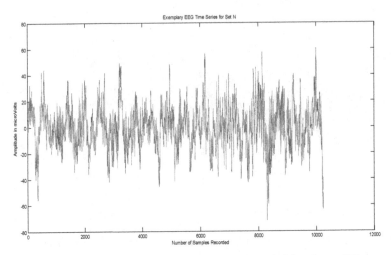

Fig. 4. Exemplary EEG time series from the recorded data for set NF.

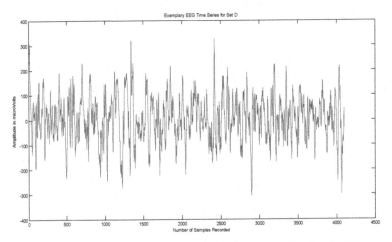

Fig. 5. Exemplary EEG time series from the recorded data for set SFI.

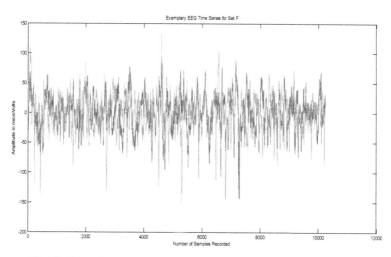

Fig. 6. Exemplary EEG time series from the recorded data for set EZF.

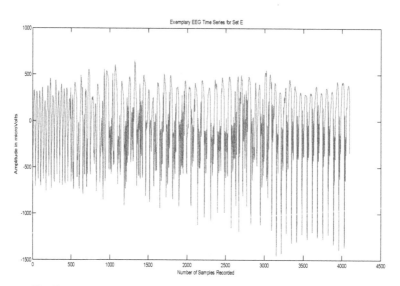

Fig. 7. Exemplary EEG time series from the recorded data for set SZR.

3.2 Statistical Analysis of EEG Signals

The statistical analysis of the dynamical properties of the EEG signals is done by applying two tests on the available datasets. First is the randomness test which is based on a nonlinear prediction error and univariate surrogates and second is the stationarity test that uses a combination of linear fluctuation measures with univariate surrogates. These tests require the state space reconstruction and evaluation of coarse-grained flow average as described in [9].

3.3 Design and Implementation of Fuzzy Inference System

It was observed that direct interpretation of the results of dynamical analysis is a tedious task and requires a lot of expertise. An attempt was hence made to develop a fuzzy inference system for decision support. A fuzzy inference model consisting of two inputs and one output using Mamdani Logic is constructed using FIS editor to categorize the datasets into different brain states. The proposed model has two inputs, rejection counts for the randomness test and the rejection counts for the stationarity test. The output gives the resultant state of the brain based on this defined input range. The generalized MATLAB R2014a fuzzy inference system is designed as in Fig. 8.

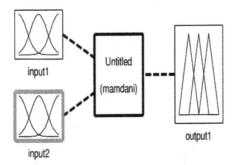

Fig. 8. Generalized fuzzy inference system (Fuzzy logic designer, MATLAB R2014a).

4 Results and Discussion

The aim of this study is to analyze and compare dynamical properties of the Electroencephalographic (EEG) signals generated from a healthy brain with those generated by the brain suffering from epilepsy and categorize them in different zones and state of the brain they belong to using a fuzzy inference model. From the healthy volunteers, Extracranial recordings recorded from the scalp of the brain have been taken. For the patients suffering from epilepsy, recordings are taken directly from the brain known as the Intracranial Electroencephalography (iEEG) or the Electrocorticography (ECoG). These have been analyzed independently by researchers for different objectives as cited in the public database. In this work, both are studied together to categorize them into different brain states they represent based on statistical analyses using Randomness test and Stationarity test.

The iEEG is divided into two parts for analysis namely the Epileptic Zone i.e. the region of the brain where epileptic activity is found and the Non Epileptic Zone which is free from the epileptic activity. Also within the epileptic zone these brain signals are analyzed during different brain states which includes the seizure free interval, the state where the first ictal activity changes is found and during the seizure activity.

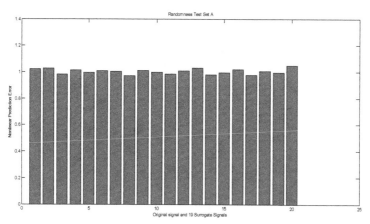

Fig. 9. Nonlinear prediction error of the original signal and the corresponding 19 surrogate signals for the Randomness test (not rejected) for set HEO.

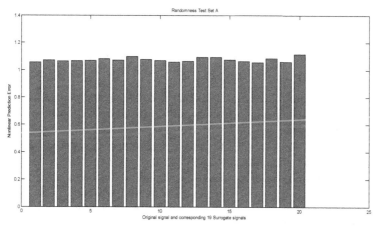

Fig. 10. Nonlinear Prediction Error of the original signal and the corresponding 19 surrogate signals for the Randomness test (rejected) for set HEO.

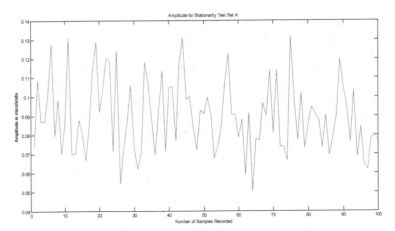

Fig. 11. Amplitude plot for the stationarity test for set HEO.

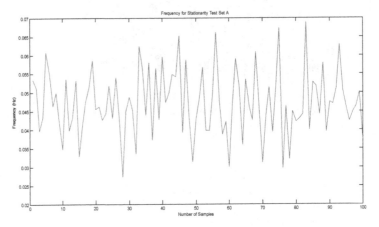

Fig. 12. Frequency plot for the stationarity test for set HEO.

Figures 9, 10, 11, and 12 presents plots for the randomness and stationarity test for the dataset HEO. The comparative analysis studied based on the rejection count does not produce any clear differentiation between the datasets belonging to different zones and states of the brain. Thus, the objective of this study was to develop a fuzzy inference model based on the rejection counts obtained from the applied tests and categorize these datasets into proper ranges which gives an implication to the brain states they belong to. The fuzzy model is constructed based on the membership function and defined rules which successfully produced and verified the results. The surface fuzzy inference model formed through the FIS editor for this rule base system is shown in Fig. 13.

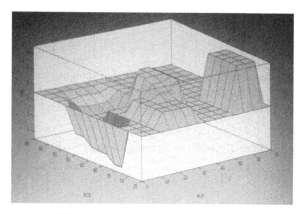

Fig. 13. Resultant surface fuzzy Model

On the basis of rules defined, a Fuzzy Rule Base was designed. RCR represents the rejection counts for the Ramdomness test. RCS represents the rejection counts for the Stationarity test and SB represents the State of Brain the EEG signal belongs to on the basis of these two input parameters.

Table 1. Definitions of input and output in the fuzzy inference model

	RCR		RCS		SB	
	Range	State	Range	State	Range	State
1	0–10	Extremely Low (EL)	25–35	Extremely Low (EL)	0–1	Healthy Eyes Opened (HEO)
2	8–20	Very Low (VL)	30–40	Very Low (VL)	1–2	Healthy Eyes Closed (HEC)
3	12–30	Low (L)	35–45	Low (L)	2–3	Non Epileptic Zone (NEZ)
4	20–35	Medium (M)	40–50	Medium (M)	3–4	Non Focal (NF)
5	25–40	High (H)	45–55	High (H)	4–5	Seizure Free Interval (SFI)
6	35–55	Very High (VH)	50–60	Very High (VH)	5–6	Epileptic Zone Focal (EZF)
7	50–70	Extremely High (EH)	55–65	Extremely High (EH)	6–7	Seizure Activity (SZR)

The concluding remarks of this study and future scope has been discussed in the next section.

5 Conclusion and Future Scope

5.1 Conclusion

On basis of the rule based analysis and fuzzy model implementation as discussed in previous section, resultant fuzzy output Table 2 is formed for representing the states of brain SB, based on the input ranges of RCS and RCR as described in Table 1.

Table 2. Output of the fuzzy inference model

RCR \ RCS	EL	VL	L	M	H	VH	EH
EL			HEO	HEO HEC	HEC		
VL					NEZ	NEZ	NF
L	SFI	SFI			NEZ	NEZ	NF
M	SFI	SFI	EZF	EZF			
H			EZF	EZF			
VH							
EH	SZR	SZR					

Table 2 is further zoned into three categories, the first zone contains extracranial recordings of the Healthy state of the brain, HEO and HEC, as marginal difference was observed between their characteristics, they belong to the same zone. Similarly, the second zone belongs to the Non Epileptic Zone from the intracranial EEG data recorded defining two different states NF and NEZ. And the last zone belongs to the Epileptic zone from the intracranial EEG data, defining three different states SFI, EZF and SZR.

However, in this resultant fuzzy inference model, it is observed that there are empty classes which shall be considered to be filled with data collected from other neurological disorders such as Schizophrenia, Parkinson's disease, etc. An attempt is in progress to analyze whether results of the applied methodology on the data from these diseases also occupy separately identifiable classes in this model or not. As there is no standardized EEG pattern that can be found for any particular brain state, this kind of computational differentiation can work in assisting the clinical studies if refined results are obtained. Nevertheless, more variations in the data are expected to make the fuzzy reasoning more difficult. This may require implication of neuro-fuzzy techniques to make the model more effective and accurate.

While dealing with more real data, it may also become difficult to visualize the distinct classes but even so it is not infeasible. Advanced mathematical and computational modeling will be required to deal with this. Different feature extraction techniques can be used, however, the best two parameters shall be selected that provides the best results, as for this study, randomness and stationarity are considered. Other future scope of this work has been discussed in the following section.

5.2 Future Scope

Recordings from the brain have always been a challenging application for nonlinear signal analysis. Problems encountered in the study of brain signals often promoted the improvement of existing nonlinear measures, the development of new measures or even new concepts such as surrogates that is used in this study. The results obtained provide further evidence that the benefit of this interdisciplinary field is mutual. The application of nonlinear signal analysis can provide valuable clinical information and has a wide scope for various applications.

There is a need to optimize the design of membership functions, fuzzy partitions and rules to make the design of Fuzzy Inference System more effective. Limited data has been used for the development work and hence the decisions may be specific. Further generalization of the rules is possible with the help of more data and inputs from neurophysiological experts. The Future work shall study how combinations of the tests used here and potential further tests can be optimized to localize brain areas where seizures originate without the necessity to observe actual seizure activity.

The relevance of this study for a broader clinical application is clearly undoubted. Nevertheless, the reported findings have to be interpreted with care because there is a lack of comparison with different properties. With an improved sensitivity and specificity of nonlinear EEG analysis and fuzzy inference techniques, broader clinical applications on a larger population of patients can be envisaged.

Acknowledgement. We are greatly thankful to Andrzejak, R. G., Schindler, K., & Rummel, C for providing their work, datasets and codes in public domain for extending the academic research.

References

1. Nidal, K., Malik, A.S.: EEG/ERP Analysis: Methods and Applications. CRC Press, Taylor & Francis Group, New York (2014)
2. Schreiber, T., Schmitz, A.: Surrogate time series. Physica D **142**(3–4), 346–382 (2000). Review Paper
3. Andrzejak, R.G., Schindler, K., Rummel, C.: Nonrandomness, nonlinear dependence, and nonstationarity of electroencephalographic recordings from epilepsy patients. Phys. Rev. E **86**(4), 046206 (2012)

4. Schreiber, T., Schmitz, A.: Improved surrogate data for nonlinearity tests. Phys. Rev. Lett. **77**(4), 635–638 (1996)
5. Andrzejak, R., Chicharro, D., Lehnertz, K., Mormann, F.: Using bivariate signal analysis to characterize the epileptic focus: the benefit of surrogates. Phys. Rev. E **83**(4), 046203 (2011)
6. Andrzejak, R., Widman, G., Lehnertz, K., Rieke, C., David, P., Elger, C.: The epileptic process as nonlinear deterministic dynamics in a stochastic environment: an evaluation on mesial temporal lobe epilepsy. Epilepsy Res. **44**(2–3), 129–140 (2001)
7. Andrzejak, R.G., Schindler, K., Rummel, C. http://ntsa.upf.edu/downloads/andrzejak-rg-schindler-k-rummel-c-2012-nonrandomness-nonlinear-dependence-and.   ntsa.upf.edu. Accessed 28 Feb 2018
8. Andrzejak, R., et.al. http://ntsa.upf.edu/downloads/andrzejak-rg-et-al-2001-indications-nonlinear-deterministic-and-finite-dimensional. ntsa.upf.edu. Accessed 28 Feb 2018
9. Kantz, H., Schreiber, T.: Nonlinear Time Series Analysis. Cambridge University Press, Cambridge (1997)

Empirical Mode Decomposition Algorithms for Classification of Single-Channel EEG Manifesting McGurk Effect

Arup Kumar Pal[1,3], Dipanjan Roy[2], G. Vinodh Kumar[2], Bipra Chatterjee[1], L. N. Sharma[3], Arpan Banerjee[2], and Cota Navin Gupta[1(✉)]

[1] Neural Engineering Lab, Department of Biosciences and Bioengineering, Indian Institute of Technology Guwahati, Guwahati, India
cngupta@iitg.ac.in
[2] Cognitive Brain Dynamics Lab, National Brain Research Centre, Gurgaon, India
[3] EMST Lab, Department of Electronics and Electrical Engineering, Indian Institute of Technology Guwahati, Guwahati, India

Abstract. Brain state classification using electroencephalography (EEG) finds applications in both clinical and non-clinical contexts, such as detecting sleep states or perceiving illusory effects during multisensory McGurk paradigm, respectively. Existing literature mostly considers recordings of EEG electrodes that cover the entire head. However, for real world applications, wearable devices that encompass just one (or a few) channels are desirable, which make the classification of EEG states even more challenging. With this as background, we applied variants of data driven Empirical Mode Decomposition (EMD) on McGurk EEG, which is an illusory perception of speech when the movement of lips does not match with the audio signal, for classifying whether the perception is affected by the visual cue or not. After applying a common pre-processing pipeline, we explored four EMD based frameworks to extract EEG features, which were classified using Random Forest. Among the four alternatives, the most effective framework decomposes the ensemble average of two classes of EEG into their respective intrinsic mode functions forming the basis on which the trials were projected to obtain features, which on classification resulted in accuracies of 63.66% using single electrode and 75.85% using three electrodes. The frequency band which plays vital role during audio-visual integration was also studied using traditional band pass filters. Of all, Gamma band was found to be the most prominent followed by alpha and beta bands which contemplates findings from previous studies.

Keywords: EMD · EEG · McGurk effect · Random Forest

A. Banerjee and C. N. Gupta—Both the senior authors spent equal time.

U. S. Tiwary and S. Chaudhury (Eds.): IHCI 2019, LNCS 11886, pp. 49–60, 2020.
https://doi.org/10.1007/978-3-030-44689-5_5

1   Introduction

Electroencephalography (EEG) is an electrophysiological technique to record the electrical signals within the brain. Although EEG signals exhibit very high temporal resolution in the order of milliseconds, they lack appreciably in spatial resolution because the neural activity detected from the surface of the skull is the summation of the excitatory and inhibitory postsynaptic potentials of billions of neurons firing simultaneously at different depths within the brain. So the signal remains contaminated with artefacts from various sources which include, but not constrained to, various physiological factors (like heart beat, respiratory pulses, eye movement, muscle movement, etc) resulting in poor Signal-to-Noise (SNR) ratio eventually making subsequent extraction of relevant information for analysis purposes troublesome. To reduce artefacts, digital filters may be applied, but with requisite precautions, as they tend to filter out the EEG activity of interest alongside and as a result contort EEG waveforms severely.

Humans get an idea of the external world by assimilating information reaching to the brain from multiple sensory organs. In case of perception of speech, along with the auditory signal, the movement of the speaker's lips plays a crucial role in influencing perception. For instance, when an auditory signal is presented alone or the visual stimulus presented along with it is congruent, then it is most likely to be correctly interpreted by the listener. But, when the same audio is presented along with a semantically incongruent visual signal, then the listener tends to hear a phoneme which is completely different from both the audio and visual cue - commonly known as the McGurk Effect [1], the rate of which is greatly affected by the relative timing of the audio and the video. In this work, we analysed the EEG recordings of 15 healthy subjects when they were presented with an audiovisual (AV) clip created by superimposing /**pa**/ as audio and lip movements generated by the utterance of /**ka**/ as visual stimuli and in most cases they confirmed of hearing /**ta**/.

Empirical Mode Decomposition (EMD) helps in decomposing a signal into different frequency bands, commonly known as Intrinsic Mode Functions (IMFs) [2]. Since the introduction of EMD by Huang et al., it has been noted to be a very effective tool for processing and analysis of various non-stationary and non-linear signals like image, climate data, seismic waves, ocean waves, EEG, etc. to name a few [3–7]. It has already been reported in literature that, the differences during unisensory and multisensory perception of AV speech stimuli gets reflected primarily in the gamma band, with some information also in the alpha and beta band [8]. Posterior superior temporal sulcus (pSTS) is the main region of human brain responsible for audio-visual integration [9]. This work focuses on building a framework to decipher McGurk effect using single channel EEG signals and empirical mode decomposition. As EMD has already been quite effective in extracting features from EEG signals in different other paradigms, we tried to implement various EMD based feature extraction techniques in this work to classify the EEG data from various subjects, recorded during an experiment manifesting McGurk effect. The feature vectors thus obtained are then classified using a non-linear classifier, Random Forest. A comparative study of the different feature extraction methods proposed has been presented here.

2   Data and Pre-processing

The AV stimulus presented to the subjects in this paradigm is created by super-imposing an audio signal of **/pa/** over the visual of lip movements formed while uttering **/ka/**, both of duration 450 ms in an audio-visual clip of 2803 ms. As the relative timing of the audio and movement of lips greatly affects the rate of cross-modal perception, the stimulus was presented in three variations:

- **−450 ms lag** where the audio signal leads the visual by 450 ms as in Fig. 1a.
- **+450 ms lag** where the audio lags visual by 450 ms as in Fig. 1b.
- **0 ms lag** where both the audio and visual stimuli are synchronous to each other as in Fig. 1c.

30 repetitions of each of the variations of AV stimulus mentioned above were taken together and randomized to form a trial block. Each subject was presented with 3 such blocks. Thus, a total of $(30 \times 3) \times 3 = 270$ AV stimuli were presented as elaborated in Fig. 1d. After each clip was played, the subjects had to confirm what they heard, from three options: */ta/*, */pa/* and */others/*. The inter-trial interval was varied somewhat between 1200 and 2800 ms to minimize expectancy effects and during that time no clip was shown on the screen.

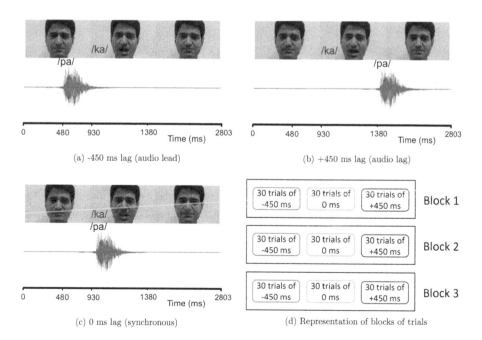

(a) -450 ms lag (audio lead)

(b) +450 ms lag (audio lag)

(c) 0 ms lag (synchronous)

(d) Representation of blocks of trials

Fig. 1. a–c: Different AV stimuli which were presented to the subjects, adapted from [10]. **d** - Pictorial representation of the trial blocks.

When the stimulus was played on the screen, simultaneous recording of EEG was done from the scalp by placing 64 electrodes in 10–20 fashion, mounted on a Neuroscan cap. Reference electrodes were linked to mastoids and grounded to AFz. As the data from 64 electrodes (known as 64 channel EEG data) was recorded at a sampling frequency of 1 kHz while the stimulus being of 2803 ms duration, each trial of EEG signal turned out to be of dimension 2803 × 64 (for 64 channels).

After referencing, EEG data from 15 subjects was band-passed through a zero phase-shift offline filter of 0.2–200 Hz. After that, a notch filter with a cut-off frequency of 50 Hz was applied on the data. Next, a baseline correction was performed on the filtered data by removing the temporal mean of the EEG signal on an epoch-by-epoch basis. Epochs with a maximum amplitude above 100 μV or minimum below −100 μV were removed to eliminate the trials contaminated with ocular and muscle-related activities [11]. On removal of the corrupted trials, we were left with the no. of trials as indicated in Table 1 for further analysis. The trials containing responses other than /ta/ and /pa/ are eliminated before pre-processing. Figure 2 depicts all the pre-processing steps described so far.

Table 1. No. of trials left for analysis after artifact rejection.

Lag ↓	Responses →	
	/ta/	/pa/
0 ms	622	200
−450 ms	458	296
+450 ms	456	288

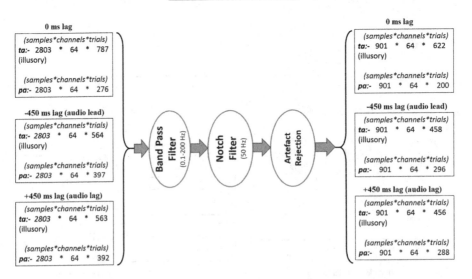

Fig. 2. A bird's eye view of the pre-processing steps applied on the EEG data.

Further details can be obtained in [8]. After artifact rejection, rest of trials are filtered using a band-pass filter having cutoff frequency of 2 Hz and 40 Hz.

3   Methodology

Four EMD based algorithms proposed here are applied on electrodes separately, i.e. neither any multidimensional approach is applied, nor the feature vectors formed from different electrodes have been concatenated. Multidimensional (taking more number of electrodes) scheme of classification showed a gradual decrease in overall accuracy with increasing no. of electrodes. So our approach was to acquire the most relevant information from the minimum number of electrodes. Reduction in the number of electrodes without compromising on the quality of information enhances the portability in case of real-time applications. Had the feature vectors from different electrodes been concatenated, then the dimension of the feature vectors would have increased for the same number of trials and thereby intensifying the *curse of dimensionality* [12]. For these reasons, the algorithms discussed in this study describe the process of obtaining feature vectors for a single electrode of a trial. Separate RFs are trained for the different electrodes in an uni-variate way. After obtaining accuracies from all the electrodes, sorting is done in a descending order of accuracy. Voting is taken from the best 3 electrodes to obtain a final label. Trail will denote the EEG signal from a single electrode of the trial, for the rest portion of the current study. Algorithm I is inspired from [7]. The temporal characteristics of cross-modal speech perception, in algorithm II to IV is conceived from [8].

3.1   Algorithm I

The algorithm described in this section has been applied to P300 paradigm in [7] and provided quite satisfactory results. The steps are discussed below:

1. All trails corresponding to /ta/ responses and /pa/ responses were ensemble averaged separately.

$$\text{pa}_{\text{avg}}(k) = \frac{1}{M_1} \sum_{i=1}^{M_1} x_{1,i}(k), \qquad \text{ta}_{\text{avg}}(k) = \frac{1}{M_2} \sum_{i=1}^{M_2} x_{2,i}(k) \qquad (1.1)$$

   where M_1 and M_2 are the number of trials corresponding to /**pa**/ $(x_{1,i}(k))$ and /**ta**/ $(x_{2,i}(k))$ respectively.
2. EMD is applied to the 2 averaged signals, obtaining their IMFs:

$$\text{pa}_{\text{avg}}(k) = \sum_{i=1}^{N_1} c_{1,i}(k), \qquad \text{ta}_{\text{avg}}(k) = \sum_{i=1}^{N_2} c_{2,i}(k) \qquad (1.2)$$

   where N_1 and N_2 are the number of IMFs obtained on decomposing pa_{avg} and ta_{avg} respectively, $c_{1,i}(k)$ and $c_{2,i}(k)$ being their respective IMFs.

3. Any trail $x(k)$ can be expanded on the basis formed by the IMFs as:

$$x(k) \simeq \sum_{i=1}^{N_1} b_{1,i} c_{1,i}(k) = \hat{x}_1(k), \qquad x(k) \simeq \sum_{i=1}^{N_2} b_{2,i} c_{2,i}(k) = \hat{x}_2(k) \qquad (1.3)$$

where $b_{1,i}$ and $b_{2,i}$ are the expansion coefficients based on the IMFs of averaged trials pa$_{\text{avg}}$ and ta$_{avg}$ respectively. The coefficients are calculated by the pseudo-inverse method and least square constraint as follows:

$$A_1 b_1 = \hat{x}_1, \qquad A_2 b_2 = \hat{x}_2 \qquad (1.4)$$

$$A_1 = \begin{bmatrix} c_{1,1}(0) & c_{1,2}(0) & \cdots & c_{1,N_1}(0) \\ c_{1,1}(1) & c_{1,2}(1) & \cdots & c_{1,N_1}(1) \\ \vdots & \vdots & \ddots & \vdots \\ c_{1,1}(K-1) & c_{1,2}(K-1) & \cdots & c_{1,N_1}(K-1) \end{bmatrix}$$

$$A_2 = \begin{bmatrix} c_{2,1}(0) & c_{2,2}(0) & \cdots & c_{2,N_2}(0) \\ c_{2,1}(1) & c_{2,2}(1) & \cdots & c_{2,N_2}(1) \\ \vdots & \vdots & \ddots & \vdots \\ c_{2,1}(K-1) & c_{2,2}(K-1) & \cdots & c_{2,N_2}(K-1) \end{bmatrix} \qquad (1.5)$$

where K is the length of the signal.

4. From A_1, A_2 and x, the vectors of expansion coefficients can be constructed as:

$$b_1 = [b_{1,1} \quad b_{1,2} \quad \cdots \quad b_{1,N_1}] = (A_1^T A_1)^{-1} A_1^T x$$
$$b_2 = [b_{2,1} \quad b_{2,2} \quad \cdots \quad b_{2,N_2}] = (A_2^T A_2)^{-1} A_2^T x \qquad (1.6)$$

5. The vectors b_1 and b_2 are concatenated together to form the feature vector.

- EEG contains a lot of time-variant noise, which may not be decomposed to a single band and get mixed with the actual signal. So, the above algorithm is also implemented replacing EMD with its noise assisted version, ICEEMDAN.

3.2   Algorithm II

Increased global gamma-band coherence and decreased alpha and beta band coherence are observed around the duration of 300–600 ms after the presentation of synchronous incongruent AV stimulus, in case of multi-sensory perception rather than unisensory perception, as reported by Vinodh et al. in [8]. It is evident that the features for classification between 2 classes will remain embedded in the spectral domain around the said temporal window. Based on this finding, instantaneous frequency feature vectors are computed using Hilbert-Huang Transform (HHT) [2].

1. A trial is divided into 3 sections: 1–300 ms, 301–600 ms and 601–900 ms.
2. Single channel EMD is applied separately on each of the three sections.
3. As in most of the trials, gamma band (30–50 Hz) is found in abundance in the 1st IMF and alpha (8–13 Hz) and beta (13–30 Hz) band frequencies are prevalent mostly in the 2nd IMF, first 2 IMFs of each section are considered.
4. Apply Hilbert Transform to the pair of IMFs obtained from each section and obtain the instantaneous frequencies.

   Now, for a particular trial, we have 6 vectors of instantaneous frequencies (*2 from each section*).
5. From the instantaneous frequency vectors of the 1st IMFs, the region of occurrence of gamma band is found out and its corresponding average timing is noted. At places, where gamma is present, corresponding energy of the signal is also obtained.
6. In the similar fashion, region of occurrence of alpha and beta bands were sought after in the 2nd IMF and the time of occurrence and average energy of alpha and beta bands are also considered as features.

In this algorithm, EMD couldn't be replaced with any of the improved versions. As EMD is applied 3 times on a particular trial, applying improved versions of EMD takes prolonged time to be executed.

3.3   Algorithm III

We noticed, EMD generates some unwanted frequencies towards the edges of a signal, when applied on a chirp signal. This problem may become prominent while performing EMD separately on three different sections after trisecting a particular trial (*as described in algorithm II*). To avoid this, and also to retain the focus on the power of different frequency bands at three different sections, EMD was applied on the trials before segmenting it and the power of different frequencies was observed at various portions of the signal. Steps of the algorithm are described below:

1. Let N be the number of IMFs that are obtained on applying EMD to the ith trial.
2. Select the 1st IMF and compute Hilbert spectrum to obtain the instantaneous frequency vector.
3. Divide the frequency vector into three sections 1:300, 301:600 and 601:900.
4. Locate the presence of gamma band in all of the three sections.
5. Note the presence (average timing) of the gamma band and it's corresponding energy in all the three sections.
6. Select the 2nd IMF and repeat steps 2 and 3.
7. Detect the alpha and beta bands from the three sections.
8. Location of occurrence (average timing) of the alpha and beta bands and their corresponding energies form the last dozen ($2 \times 2 \times 3 = 12$) of features of the feature vector.

In the above algorithm also, updated versions of EMD couldn't be applied on the trials because of the same reason of high computation time.

3.4   Algorithm IV

Although most of the information for multi-sensory speech integration of synchronous AV stimuli is found to accumulate within the top three frequency bands, no such finding was reported in the literature for stimuli with lags, both for positive and negative. Keeping this in consideration, this algorithm is proposed so that it can be applied on EEG data for synchronous as well as asynchronous AV stimuli. Here also, only the first 2 IMFs are considered as many trials failed to generate more than a pair. The steps are documented below:

1. Apply EMD to a particular trial and select the 1st pair of 'oscillatory modes'.
2. Concatenate the IMFs one after another to form the feature vector.

4   Results

Features generated from the proposed algorithms, are classified as depicted in Fig. 3 and labels from the top three electrodes exhibiting maximum accuracy are taken into consideration for prediction of the final label of the test data. Like most of the non-linear classifiers, random forest also requires fine tuning of a few parameters, like increase in the number of trees causes a rise in the accuracy (*though at the cost of computational complexity and time*) which saturates after reaching a certain number. In this case, the accuracy gets saturated when the number of trees in the RF reaches 200. A comparative study among the classification accuracies of different feature extraction techniques, presented in Table 2, reveals algorithm I to be more effective than others in all of the 3 cases of AV lags.

It turns out that, owing to the very high dimension of the feature vectors in algorithm IV, it took considerable amount of time in classifying the test data but the accuracy didn't improve appreciably. Whereas, algorithm I when modified with Improved Complete Ensemble EMD (ICEEMDAN) [13], the time required for classifying with this enhanced version of EMD was not much high with respect to the satisfactory classification accuracy it achieved. The parameters chosen for the training of the Random Forest classifiers in this study is as follows [14]:

– No. of samples for constructing a decision tree:- a randomly sampled number between 100 and 150.
– No. of trees in the forest:- 200.

The parameters for ICEEMDAN are:

– β_0 is kept constant at 0.2.
– Number of realizations of noise:- 800.
– Maximum number of iterations before terminating:- 200.

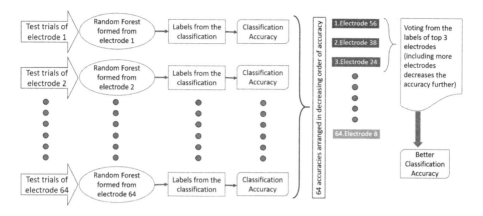

Fig. 3. Feature vectors of individual electrodes are classified through separate random forests and top 3 are considered for final labels.

4.1   Most Informative Frequency Band

This study also aims at finding the frequency band which plays the most vital role in audio-visual integration. EMD, being an entirely data-driven approach, can't guarantee the presence of a particular frequency band in a specific IMF. So, for extracting distinct frequency bands from the combination of a wide range of frequencies, we applied the traditional filter bank technique.

EEG signal from a particular trial is passed through 5 band-pass filters yielding filtered signals corresponding to 5 different frequency bands namely: gamma (30 Hz to 50 Hz), beta (13 Hz–30 Hz), alpha (8 Hz–13 Hz), theta (4 Hz–7 Hz) and delta (0.5 Hz–4 Hz) bands. After that, each trial is represented by $64 \times 5 = 320$ vectors (5 frequency bands for each of the 64 electrodes). Classification is done on these vectors separately and arranged in the order of decreasing accuracy as shown in Fig. 4. On completion of that, starting from the band of the highest accuracy, groups of 10 are formed with decreasing accuracy i.e. 10th member of group 32 will have the least accuracy. In each of these 32 groups, the number

Table 2. Classification accuracies for different feature extraction techniques, on being classified with RF containing 200 trees each.

	Single electrode			Top 3 electrodes		
AV lags →	−450 ms	0 ms	+450 ms	−450 ms	0 ms	+450 ms
Algorithm I (EMD)	**62.8**	63.66	**60.14**	**66.13**	75.85	**64.59**
Algorithm I (ICEEMDAN)	58.27	**65.37**	59.59	64.13	**76.95**	63.92
Algorithm II	59.87	62.56	57.30	62.27	75.37	61.35
Algorithm III	58.4	**65.37**	54.05	58.67	74.75	61.08
Algorithm IV	56.53	64.39	56.22	62.4	72.20	60.68

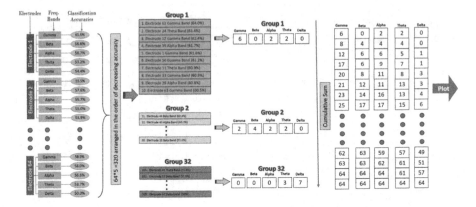

Fig. 4. The figure depicts the flow diagram for finding the relative activeness of different frequency bands. The data in the above figure is obtained when EEG signal for synchronous AV stimuli is classified with RF containing 200 trees each.

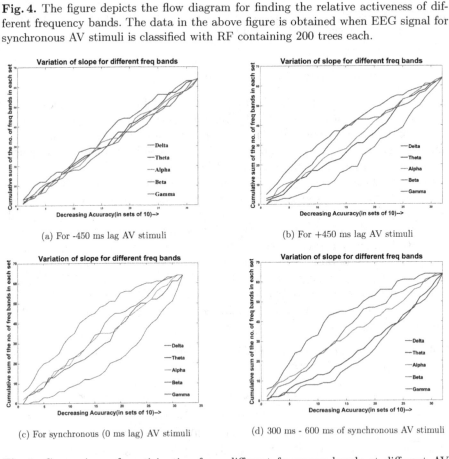

(a) For -450 ms lag AV stimuli

(b) For +450 ms lag AV stimuli

(c) For synchronous (0 ms lag) AV stimuli

(d) 300 ms - 600 ms of synchronous AV stimuli

Fig. 5. Comparison of participation from different frequency bands at different AV stimuli.

of gamma, beta, alpha, theta and delta bands are noted separately. After that, a cumulative sum is performed for each of these frequency bands and plotted against group no.s.

It is evident from Fig. 5c that, for synchronous AV stimuli, gamma band holds the maximum information followed by alpha and beta bands. This pattern becomes more prominent when instead of all the 901 samples, only the data from 300 ms to 600 ms is considered (Refer Fig. 5d). Both these findings are in accordance with the previous findings of [8]. The literature also suggested that no frequency bands were found to be exceptionally active with respect to others during the presentation of asynchronous AV stimuli. This is shown in by Fig. 5a and 5b.

5   Conclusion

Obtaining classification accuracy of around 60% and 70% from a single channel and three channel EEG data respectively will lead to design a fast and practical way of classifying EEG signals with McGurk paradigm. In this study, all the subjects' data were taken together to train the classifier, which overcomes the constraints of subject-specific nature. EMD feature extraction of single channel or few channels EEG produces lesser number of IMFs and dimension of the data gets reduced. Since we have knowledge about the brain regions susceptible to event-related potentials, this proposed framework may be useful to implement fast real-time ERP BCI. Also, McGurk effect has been applied to track differences in audio-visual integration between individuals. BCIs based on McGurk effect can possibly assist hearing-impaired individuals, by exploiting the integration of auditory and visual senses. This approach can also be extended to other paradigms in order to reduce the amount of data needed to reach high precision as well as facilitate the set up with a less bulky hardware.

Acknowledgement. NBRC authors (DR, GVK and AB) collected the EEG data and performed preprocessing steps. The EMD frameworks were formulated by AKP and CNG. The code for Random Forest was written in MATLAB by BC, AKP and CNG. The paper was written by AKP, BC and CNG. All authors assisted in answering the reviewers comments.

This initial part of this research was funded by NBRC core and the grants Ramalingaswami fellowship (BT/RLF/Re-entry/31/2011) and Innovative Young Biotechnologist Award (IYBA) (BT/07/IYBA/2013) from the Department of Biotechnology (DBT), Ministry of Science Technology, Government of India to Arpan Banerjee. Dipanjan Roy was supported by the Ramalingaswami fellowship (BT/RLF/Re-entry/07/2014) and DST extramural grant (SR/CSRI/21/2016). Neural Engineering Lab, IIT Guwahati is supported by Startup Grant, IIT Guwahati and NECBH grant sponsored by DBT (NECBH/2019-20/177).

References

1. McGurk, H., MacDonald, J.: Hearing lips and seeing voices. Nature **264**(5588), 746 (1976)

2. Huang, N.E., et al.: The empirical mode decomposition and the Hilbert spectrum for nonlinear and non-stationary time series analysis. Proc. R. Soc. Lond. A **454**(1971), 903–995 (1998)
3. Molla, M., Islam, K., Rahman, M.S., Sumi, A., Banik, P.: Empirical mode decomposition analysis of climate changes with special reference to rainfall data. Discrete Dyn. Nature Soc. **2006** (2006)
4. Loh, C.-H., Wu, T.-C., Huang, N.E.: Application of the empirical mode decomposition-Hilbert spectrum method to identify near-fault ground-motion characteristics and structural responses. Bull. Seismol. Soc. Am. **91**(5), 1339–1357 (2001)
5. Looney, D., Mandic, D.P.: Multiscale image fusion using complex extensions of EMD. IEEE Trans. Signal Process. **57**(4), 1626–1630 (2009)
6. Veltcheva, A.D.: Wave and group transformation by a Hilbert spectrum. Coast. Eng. J. **44**(4), 283–300 (2002)
7. Arasteh, A., Moradi, M.H., Janghorbani, A.: A novel method based on empirical mode decomposition for P300-based detection of deception. IEEE Trans. Inf. Forensics Secur. **11**(11), 2584–2593 (2016)
8. Kumar, G.V., Halder, T., Jaiswal, A.K., Mukherjee, A., Roy, D., Banerjee, A.: Large scale functional brain networks underlying temporal integration of audiovisual speech perception: an EEG study. Front. Psychol. **7**, 1558 (2016)
9. Hocking, J., Price, C.J.: The role of the posterior superior temporal sulcus in audiovisual processing. Cereb. Cortex **18**(10), 2439–2449 (2008)
10. Kumar, V.G., Dutta, S., Talwar, S., Roy, D., Banerjee, A.: Neurodynamic explanation of inter-individual and inter-trial variability in cross-modal perception. bioRxiv, p. 286609 (2018)
11. Teplan, M., et al.: Fundamentals of EEG measurement. Meas. Sci. Rev. **2**(2), 1–11 (2002)
12. Lotte, F., Congedo, M., Lécuyer, A., Lamarche, F., Arnaldi, B.: A review of classification algorithms for EEG-based brain-computer interfaces. J. Neural Eng. **4**(2), R1 (2007)
13. Colominas, M.A., Schlotthauer, G., Torres, M.E.: Improved complete ensemble EMD: a suitable tool for biomedical signal processing. Biomed. Signal Process. Control **14**, 19–29 (2014)
14. Breiman, L.: Random forests. Mach. Learn. **45**(1), 5–32 (2001)

Natural Language, Speech and Dialogue Processing

Abstractive Text Summarization Using Enhanced Attention Model

Rajendra Kumar Roul[1]([✉])[iD], Pratik Madhav Joshi[2][iD],
and Jajati Keshari Sahoo[3][iD]

[1] Department of Computer Science, Thapar Institute of Engineering and Technology,
Patiala 147004, Punjab, India
`raj.roul@thapar.edu`
[2] Department of Computer Science, BITS-Pilani, K. K. Birla Goa Campus,
Zuarinagar 403726, Goa, India
`pratikmjoshi123@gmail.com`
[3] Department of Mathematics, BITS-Pilani, K. K. Birla Goa Campus,
Zuarinagar 403726, Goa, India
`jksahoo@goa.bits-pilani.ac.in`

Abstract. Text summarization is the technique for generating a concise
and precise summary of voluminous texts while focusing on the sections
that convey useful information, and without losing the overall meaning.
Although recent works are paying attentions in this domain, but still they
have many limitations which need to be address.. This paper studies the
attention model for abstractive text summarization and proposes three
models named as *Model 1*, *Model 2*, and *Model 3* which not only handle
the limitations of the previous contemporary works but also strengthen
the experimental results further. Empirical results on CNN/DailyMail
dataset show that the proposed approach is promising.

Keywords: Abstractive · Attention · Encoder · ROUGE ·
Transformer

1 Introduction

Text summarization refers to the technique of shortening long pieces of text.
The intention is to create a coherent and fluent summary having only the main
points outlined in the document. Text summarization is broadly categorized into
two types- *extractive* and *abstractive*. The abstractive text summarization algorithms create new phrases and sentences that relay the most useful information
from the original text – just like humans do. The abstraction technique entails
paraphrasing and shortening parts of the source document. When abstraction is
applied for text summarization in deep learning problems, it can overcome the

U. S. Tiwary and S. Chaudhury (Eds.): IHCI 2019, LNCS 11886, pp. 63–76, 2020.
https://doi.org/10.1007/978-3-030-44689-5_6

grammar inconsistencies of the extractive method. On the other hand, extractive text summarization technique involves pulling keyphrases from the source document and combining them to make a summary. The extraction is made according to the defined metric without making any changes to the texts. The abstractive mechanism is more challenging than extractive due to the involvement of complex neural network based architectures such as recurrent neural network (RNN) and Long short-term memory (LSTM). Naturally abstractive approaches are harder. For perfect abstractive summary, the model has to first truly understand the document and then try to express that understanding in short possibly using new words and phrases. Has complex capabilities like generalization, paraphrasing and incorporating real-world knowledge. The corpus is compressed in a lossy manner while maintaining the meaning and salient features of the original data. Many research works have been done in the field of extractive text summarization [1–4], however innovative work in the domain of multi-document text summarization using abstractive technique is limited [5–8].

Aiming in this direction, this paper studies the attention mechanism used for abstractive text summarization and develops three modified techniques using attentional encoder-decoder recurrent neural networks that improves the abstractive summary of a large corpus having multiple documents. The first modified technique named as 'Model 1' is based on *Specialized Sequence-to-Sequence Architecture*. The second and third modified techniques named as 'Model 2' and 'Model 3' are based on *Transformer Architecture* and *Multi-task Learning Network Architecture* respectively. Experimental results on CNN/DailyMail dataset show that the proposed techniques are better than the existing state-of-the-art approaches.

2   Preliminaries

2.1   Attention Model

The attention mechanism in neural networks is not a particularly modern concept, especially in image recognition. However, recent studies show that this mechanism has found its large scope of application in natural language processing [9]. Perhaps the best way to understand the scope of the attention model in natural language processing would be to take neural machine translation (NMT) as an example. Traditional machine translation systems typically rely on using the inherent properties of text to generate sophisticated features by clever feature engineering. A lot of engineering effort is devoted in order to create and fine-tune these complex systems. NMT works a bit differently where the meaning of a sentence is mapped into a fixed-sized representational vector which is then used to generate a translation. By not relying on things like n-gram counting and instead trying to capture the higher-level meaning of a text, NMT systems are much more effective at generalizing to new sentences than the existing systems.

2.2 Attention Model for Machine Translation

NMT uses an encoder-decoder architecture that encodes a source sentence into a fixed-length vector [6]. Using this, the decoder generates a translation mechanism as illustrated in Fig. 1. Fixed-length vector is the primary bottleneck in improving the performance of the novel encoder-decoder architecture system, and our paper proposes to improve upon this by allowing a model to automatically search for parts of the source sentence that are important for predicting a target word without having to explicitly creating a hard segment to form these parts. This new approach is very effective and yields results comparable to the existing state-of-the-art approaches previously used for English-to-French translation. In our investigation, we found that this model can be effective when applied to the problem of summarization as well. The model has been used in many state-of-the-art works with several modifications [5,6].

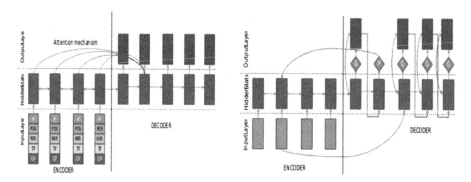

Fig. 1. Attention model for machine translation.

Fig. 2. Switching generator-pointer.

2.3 Using a Feature-Rich Encoder

This method aims to add more meaning and features to the embeddings generated by the encoder and also solve the issue of prediction of named entities. To do this, the conventional word embeddings are combined with the additional linguistic features such as part-of-speech(POS) tags, named-entity tags, and TF-IDF statistics of the words. They combine to form a modified embedding which is more rich in features [10].

2.4 Modeling on Unseen/Rare Keywords Using Switching Generator-Pointer

This novel method is an improvement upon previously tried methods such as using 'UNK' tokens (represents an unknown word) as placeholders, the most

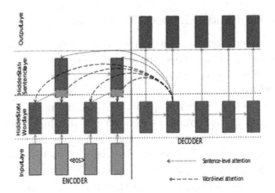

Fig. 3. Hierarchical attention model.

common methodology for handling rare/unseen words. This method involves of creating a switch that is modeled as a sigmoid activation function over a linear layer which points to the appropriate position of the rare/unseen word in the source document as illustrated in Fig. 2. In order to pinpoint important key-words as well as important sentences for the summary, a combined hierarchical attention layer is used to incorporate attention at both levels. This enhances the focus of the hierarchical nature of documents and generates a net attention over the document as shown in Fig. 3.

3   Proposed Approach

3.1   Generalized Sequence-to-Sequence Model

Consider a corpus C containing documents $d = \{d_1, d_2, \cdots, d_n\}$. The golden multi-sentence summaries $s = \{s_1, s_2, \cdots, s_n\}$ are generated by human anno-tation, which often can be done through crowd sourcing or export annota-tion [11]. The proposed approach uses export-annotated golden multi-sentence summaries for the experimental work. The proposed generalized sequence-to-sequence model M treats the summarization problem as that of machine trans-lation and it consists of an encoder E and a decoder D. Essentially, for each document $d_i = \{w_1^i, w_2^i, ..., w_k^i\}$, with k words, the encoder E encodes the docu-ment word by word to a latent vector representation L_i. Using L_i, the decoder D generates summary $\hat{s}_i = \{\hat{w}_1^i, \hat{w}_2^i, \cdots, \hat{w}_j^i\}$ with j words, word by word[1]. A depic-

---

[1] Here $\hat{w}_j^i$ isn't an estimate of the word w_j^i, but a prediction of the jth word of the summary.

tion is illustrated in Fig. 4 where the source document words $X_1, X_2, \cdots, X_T$ are converted to latent representation of C and decoded to $y_1, y_2, \cdots, y_T$.

3.2 Preprocessing of Documents

The following steps are involved in the preprocessing of the documents of the corpus C.

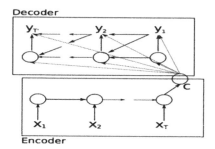

Fig. 4. Sequence-to-sequence model.

1. *The source documents and target summaries are tokenized:*
   Since we are given raw text documents and summaries, we need to convert the text to an array of *tokens*, which include words and punctuations. NLTK word tokenizer[2] is used to generate the word arrays. This tokenizer is better than a conventional space separation tokenizer as it draws from a dictionary to recognize words, and also splits with respect to punctuation and apostrophes.
2. *Source document and golden summary length is controlled via truncation and padding:*
   In order to speed up training, we keep a constant source document and summary size, and limited the number of tokens to N_d (number of tokens per document) and N_s (number of tokens per summary) respectively. We also pad shorter documents and summaries to a designated number of tokens in order to ensure consistency.
3. *Limit vocabulary size to prevent large vector size:*
   The decoding speed is boosted by limiting the shared vocabulary of the input documents and summaries to a common number N_{vocab}. Any additional words are replaced by an UNK token. However, this can sometimes slow down the performance of the model. Therefore, we have applied this step to only '*Model 1*' and '*Model 2*' and not to '*Model 3*' (as introduced in Sect. 1).

---

[2] https://www.nltk.org/api/nltk.tokenize.html.

4. *Vectorizing of the tokens by using pretrained GloVe vectors:*
   All the tokens are converted into vectors so that the model can process them. A standard method is to convert unique token (word) w_k into a one-hot vector (a standard term used in machine learning for vector) $V_k = [0, 0, \cdots, 1, \cdots, 0]$, where all values except the kth position are 0, and k corresponds to the index of the unique token in the vocabulary. However, in order to improve the "understanding" of the model, we incorporate the background knowledge by using pretrained GloVe vectors[3]. This is more beneficial than simply one-hot encoding of the words, because not only does the model have a greater understanding of the relationship between tokens (thus allowing it to better comprehend syntactic and semantic nuances in the text), but the task is also sped up by relieving the model of understanding the same.

3.3   Architecture of the Proposed Models

Three models are proposed, which are all modifications of the vanilla sequence-to-sequence (seq2seq) architectures as approaches to the summarization task and are discussed below.

Model 1: Specialized Sequence-to-Sequence Architecture:- A modified seq2seq architecture [6] is used which contains a global attention mechanism. This mechanism is proposed as a solution to the shortcomings of the vanilla seq2seq architecture. In a simple encoder-decoder model, long sequences of tokens are difficult to decode, since the latent representation L is a fixed vector and may not give sufficient information about the model. Global attention does the task of both alignment and translation. In addition to this, the proposed approach used the concept of copy attention [12] when dealing with documents which are filled with names of entities and complex and rare terminologies. The standard attention mechanism is reused for the copy attention mechanism in order to prevent a large number of trained parameters. LSTM cells are used as opposed to vanilla RNN cells and a bidirectional RNN [13] is used for the encoder in order to gain a more enriched encoding of the input text.

Model 2: Transformer Architecture:- The concept of the transformer architecture is inspired from [14]. This model although is a seq2seq architecture, does not use the concept of recurrent neural networks, but rather a self-attention mechanism. RNNs are computed and processed in a sequential manner making parallelism difficult for such models. The proposed transformer architecture uses a self-attention mechanism which models the relationships between all the tokens (words) in a sentence and generates an attention score. These scores then decide

---

[3]   https://nlp.stanford.edu/projects/glove/.

the contribution to the representation of the next word. The self attention mechanism consists of multi-headed attention layers. The model consists of stacked feedforward layers and the multi-headed attention layers. An illustration of the model and multi-headed attention layers are shown in Figs. 5 and 6 respectively. These layers allow parallelism and largely speed up the training which allow us to train the model thoroughly in an efficient manner.

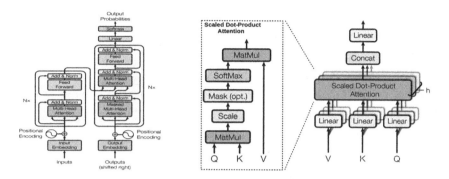

Fig. 5. Transformer architecture.   **Fig. 6.** Multi-headed attention layers.

Model 3: Multi-task Learning Network Architecture:- Consider MLM be a general multi-task learning model which trains on a set of tasks $T = \{t_1, t_2, \cdots, t_n\}$. These tasks need to be related in the sense that they have to bring similar changes to a loss function, not opposing changes. An improvement in performance of one task shouldn't degrade the performance of another. MLM trains these tasks on a common loss function J_{MLM}. This leads to an inherent transfer of knowledge between various tasks and the joint training can lead to an increased performance across all the tasks.

The proposed approach employs a fusion of the two models, the seq2seq model and multi-task learning model. For this, the Multitask Question Answering Network (MQAN) [15] has been used. It is a large, complex, and generalized model which integrates techniques like self-attention [14] and co-attention to generate high performance results across a variety of natural language problems. It does so by treating each task as that of a question-answering nature with context. The model architecture is shown in Fig. 7. The set of tasks T jointly trained are summarization, English-to-German machine translation using the IWSLT dataset [16], natural language inference using the *multiNLI* dataset [17], question-answering using the *SQuAD* dataset [18], semantic role labeling using the *QA-SRL* dataset [19], semantic analysis using the *Stanford Sentiment Treebank* [20], semantic parsing using the *WikiSQL* dataset [21], goal-oriented dialogue using the *WOZ* dataset [22], and zero-shot relation extraction using the *ZRE* dataset [23].

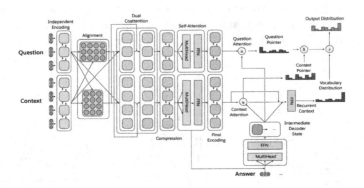

Fig. 7. Architecture of multitask question answering network.

4 Experimental Results

4.1 CNN/DailyMail Corpus

For experimental purposes, CNN/DailyMail (CNN/DM) [5] dataset has been used which contains online news articles paired with golden multi-sentence summaries. The statistics of CNN/DM dataset is shown in Table 1. We used the non-anonymized, cleaned version as used in [12]. 5000 data points (or documents) are randomly sampled from the CNN/DM dataset to generate the corpus. The source documents and the target summaries are truncated to $N_d = 400$ tokens (words), $N_s = 100$ tokens (words) respectively. The vocabulary for decoding was limited to $N_{vocab} = 500$ words (except for MQAN). The entire corpus is divided into 3750 data points of training, and 625 each for validation and testing. The size of pretrained GloVe embeddings is fixed to 200. The Recall-Oriented

Table 1. CNN/DailyMail statistics

Information	Numbers
Avg. number of sentences per source article	29.74
Avg. number of sentences per summary	3.75
Avg. number of words per source article	766
Avg. number of words per summary	53
No. of data points (Training)	287,226
No. of data points (Validation)	13,368
No. of data points (Testing)	11,490

Understudy for Gisting Evaluation (ROUGE) score [24] is used for performance evaluation. ROUGE-N score measures the unigram, bigram, and higher order n-gram overlap between generated and gold summary. ROUGE-1, ROUGE-2, and ROUGE-L scores are used for measuring the evaluation, where ROUGE-L incorporates sentence level similarity and measures longest co-occurring in-sequence n-grams.

4.2   Experimental Setup and Model Parameters

The following parameters are used for training the proposed models where the parameter values are decided by the experiment on which better results are obtained.

1. *Seq2seq Architecture:* We set the size of encoder and decoder to 512 nodes. Stochastic gradient descent was used as the optimizer and the gradient vector is renormalized every time when its norm exceeds 3. The training continued for 10000 steps.
2. *Transformer Architecture:* For this, 8 transformer heads are used. Dropout was kept to 0.1 (as opposed to original model [14] whose dropout was 0.2) and training was done for 100000 steps. All Other parameters are similar to original model.
3. *Multi-task Learning Network Architecture:* Here, the vocabulary is considered to be its maximum size and it expanded to 785016 tokens. However the decoder vocabulary was limited to 50000 tokens. The input embeddings were pretrained GloVe vectors of size 200. Maximum output length was limited to 100 tokens and maximum context length (or in other words, the source document length) was limited to 400 tokens. Dropout was removed for this and 3 transformer heads are used. Transformer hidden layer was of size 150 and there are 2 transformer layers.

4.3   Comparison Analysis with the Competitive Models

The performance of the proposed models are compared with the following state-of-the-art models:

1. TexSum+Glove Model: This model was used in [25] that uses the open-source "TexSum" Tensorflow model, which is an encoder-decoder model with a bidirectional RNN using LSTM cells as the encoder, and a one-directional LSTM RNN with attention and beam search as the decoder. It also uses pretrained GloVe embeddings to generate input vectors. Instead of using standard attention [6] for the copy mechanism, it uses a probabilistic model to determine whether to generate the word using the decoder or simply copy the token from the source text.
2. Words-lvt5k-1sent: This model was used in [5] and uses a variety of modifications to yield advanced results while handling the computational bottleneck. These modifications make it the most competitive model. The modifications involve the large vocabulary trick, a feature-rich encoder, and the hierarchical attention.

3. ABS+: This model was used in [26]. It is a conditional language model which uses a deep attention-based encoder and beam search decoder.
4. RAS-Elman: This model was used in [27] and consists of a deep convolutional attention-based encoder and a recurrent decoder.

4.4 Discussion

The experimental results over the entire CNN/DM dataset of the three proposed modified models are compared with the state-of-the-art architectures and are shown in Table 2 where the bold result shows the highest score received by the corresponding model. From the results one can see that the proposed three models outperforms the state-of-the-art models except for some results where the performance of the proposed techniques are at par with existing techniques. Training the proposed models for more steps would probably results in the performance tending to higher ROUGE scores. The multi-task learning network had a superior performance. It performed competitively with the other known models and had far higher ROUGE scores than the other models we experimented on ROUGE-1, ROUGE-2, and ROUGE-L scores of 10 randomly selected documents for the 3 proposed models are shown in Tables 3, 4, and 5 respectively.

Table 2. Results comparison

Model	ROUGE-1	ROUGE-2	ROUGE-L
words-lvt5k-1sent [5]	0.3510	0.1723	0.3321
RAS-Elman [27]	0.3454	0.1667	0.3145
TexSum+GloVe [25]	0.1267	0.0489	0.1185
ABS+ [26]	0.3043	0.1756	0.2998
Seq2Seq architecture	0.3578	**0.1896**	0.2687
Transformer architecture	0.2623	0.1878	0.3354
Multi-task learning network	**0.3676**	0.1344	**0.3485**

For experimental purposes, we have shown the generated summary of the transformer architecture in Table 6 and multi-task learning network architecture in Table 7.

It can be seen from Tables 6 and 7 that the transformer architecture still isn't able to generate coherent, grammatically accurate sentences, whereas the multi-task learning network not only does that, it also uses a diverse vocabulary and has a complex sentence structure understanding. Where the transformer architecture doesn't complete the sentence often, the multi-task learning network maintains structural integrity and produces realistic summaries which are hard to distinguish from the actual summaries.

Table 3. Seq2Seq architecture

Document-ID	ROUGE-1	ROUGE-2	ROUGE-L
103	0.4527	0.2266	0.1842
145	0.3902	0.2032	0.2502
162	0.3411	0.2579	0.2089
288	0.2052	0.0051	0.2811
296	0.3509	0.1011	0.2205
370	0.2845	0.1005	0.1645
415	0.1213	0.0031	0.1869
447	0.2888	0.1004	0.2975
504	0.3692	0.2006	0.3553
559	0.2034	0.1002	0.2132
Average	0.2907	0.1299	0.2362

Table 4. Transformer architecture

Document-ID	ROUGE-1	ROUGE-2	ROUGE-L
103	0.2578	0.1012	0.2033
145	0.3529	0.2714	0.3623
162	0.3044	0.1428	0.3352
288	0.1818	0.1102	0.1352
296	0.2162	0.1314	0.2636
370	0.2428	0.1004	0.2676
415	0.1845	0.0802	0.1252
447	0.1764	0.0760	0.2001
504	0.0875	0.0079	0.0857
559	0.1379	0.0614	0.1534
Average	0.2142	0.1083	0.2132

Table 5. Multi-task learning network architecture

Document-ID	ROUGE-1	ROUGE-2	ROUGE-L
103	0.3876	0.2032	0.3992
145	0.4113	0.2542	0.4001
162	0.2566	0.1211	0.2206
288	0.2009	0.1094	0.2100
296	0.1821	0.1031	0.1205
370	0.2444	0.1155	0.2317
415	0.2013	0.1082	0.2666
447	0.2833	0.1677	0.2912
504	0.4774	0.2999	0.4732
559	0.3092	0.2051	0.3102
Average	0.2954	0.1083	0.2923

Table 6. Sample generated summary of transformer architecture

Target summary	Predicted summary
Border Control agents have allegedly witnessed gang members recruiting members from among the influx of unaccompanied immigrant minors from Central America. Shawn Moran of the National Border Control Patrol Council says agents have told him about the enlistment. Border Control agents have allegedly witnessed gang members using a Red Cross phone bank at the facility at Nogales, Arizona, to 'recruit, enlist and pressure' other minors. Authorities say there is no evidence of such activity	The vice illegally illegally for than The to southwestern men and said October are sent so week of than said and said young young they policies who in some

Table 7. Sample generated summaries of multi-task learning network architecture

Target summary	Predicted summary
officials: agent fired when a suspected illegal border crosser assaulted him. incident happened along u.s.-canada border	the agent fired when a suspected "illegal border crosser" assaulted him during a patrol along the u.s.-canadian border. the agent was struck in the head
tropical cyclone pam struck the island nation of vanuatu. aid workers on the ground are reporting massive destruction	tropical cyclone pam pummeled the south pacific island nation of vanuatu. the government declared a state of emergency as the death toll continues to rise
plainclothes officers were wounded by gunfire. la times: police believe they were targeted. lapd pio:"this cannot be called targeting"	two individuals were detained for questioning. the officers were on duty in south l.a., when gunfire erupted in the early evening. the officers were on duty in south l.a. when gunfire erupted in the early evening

5   Conclusion

The problem of abstractive text summarization is an unsolved problem that has many obstacles. There is a prominent issue of human readable summaries to use as target data. By effectively implementing the optimizations techniques in the existing works, we can possibly emulate the state-of-the-art performance given that we have extended computational resources for the same. This paper studied the attention model for abstractive text summarization and proposed three modified techniques to further strengthen the summary. Results on CNN/DM corpus show that the proposed techniques can give better performance compared to other well established recent techniques. A possible improvement of the proposed approach could be to generate an ensemble model which takes into account multiple neural models and generates outputs based on a majority vote of such models. Also, this work can be extended by combining the extractive mechanism along with the abstractive technique for text summarization.

References

1. Roul, R.K., Sahoo, J.K., Goel, R.: Deep learning in the domain of multi-document text summarization. In: Shankar, B.U., Ghosh, K., Mandal, D.P., Ray, S.S., Zhang, D., Pal, S.K. (eds.) PReMI 2017. LNCS, vol. 10597, pp. 575–581. Springer, Cham (2017). https://doi.org/10.1007/978-3-319-69900-4_73
2. Roul, R.K., Mehrotra, S., Pungaliya, Y., Sahoo, J.K.: A new automatic multi-document text summarization using topic modeling. In: Fahrnberger, G., Gopinathan, S., Parida, L. (eds.) ICDCIT 2019. LNCS, vol. 11319, pp. 212–221. Springer, Cham (2019). https://doi.org/10.1007/978-3-030-05366-6_17

3. Roul, R.K., Sahoo, J.K.: Sentiment analysis and extractive summarization based recommendation system. In: Behera, H.S., Nayak, J., Naik, B., Pelusi, D. (eds.) Computational Intelligence in Data Mining. AISC, vol. 990, pp. 473–487. Springer, Singapore (2020). https://doi.org/10.1007/978-981-13-8676-3_41

4. Roul, R.K., Arora, K.: A nifty review to text summarization-based recommendation system for electronic products. Soft Comput. **23**(24), 13183–13204 (2019). https://doi.org/10.1007/s00500-019-03861-3

5. Nallapati, R., Zhou, B., dos Santos, C., Gulcehre, C., Xiang, B.: Abstractive text summarization using sequence-to-sequence RNNS and beyond. In: Proceedings of The 20th SIGNLL Conference on Computational Natural Language Learning, pp. 280–290. Association for Computational Linguistics (2016)

6. Bahdanau, D., Cho, K., Bengio, Y.: Neural machine translation by jointly learning to align and translate. CoRR, vol. abs/1409.0473 (2014)

7. Liu, L., Lu, Y., Yang, M., Qu, Q., Zhu, J., Li, H.: Generative adversarial network for abstractive text summarization. In: Thirty-Second AAAI Conference on Artificial Intelligence (2018)

8. Gerani, S., Carenini, G., Ng, R.T.: Modeling content and structure for abstractive review summarization. Comput. Speech Lang. **53**, 302–331 (2019)

9. Song, S., Huang, H., Ruan, T.: Abstractive text summarization using LSTM-CNN based deep learning. Multimed. Tools Appl. **78**(1), 857–875 (2018). https://doi.org/10.1007/s11042-018-5749-3

10. Roul, R.K, Bhalla, A., Srivastava, A.: Commonality-rarity score computation: a novel feature selection technique using extended feature space of ELM for text classification. In: Proceedings of the 8th Annual Meeting of the Forum on Information Retrieval Evaluation, pp. 37–41. ACM (2016)

11. Hermann, K.M., et al.: Teaching machines to read and comprehend. In: Advances in Neural Information Processing Systems, pp. 1693–1701 (2015)

12. See, A., Liu, P.J., Manning, C.D.: Get to the point: summarization with pointer-generator networks. In: Proceedings of the 55th Annual Meeting of the Association for Computational Linguistics, pp. 1073–1083. Association for Computational Linguistics (2017)

13. Schuster, M., Paliwal, K.K.: Bidirectional recurrent neural networks. IEEE Trans. Signal Process. **45**(11), 2673–2681 (1997)

14. Vaswani, A., et al.: Attention is all you need. In: Advances in Neural Information Processing Systems, pp. 5998–6008 (2017)

15. McCann, B., Keskar, N.S., Xiong, C., Socher, R.: The natural language Decathlon: multitask learning as question answering. CoRR, vol. abs/1806.08730 (2018)

16. Cettolo, M., Girardi, C., Federico, M.: Wit[3]: web inventory of transcribed and translated talks. In: Proceedings of the 16th Conference of the European Association for Machine Translation (EAMT), May, Italy, Trento, pp. 261–268 (2012)

17. Williams, A., Nangia, N., Bowman, S.: A broad-coverage challenge corpus for sentence understanding through inference. In: Proceedings of the 2018 Conference of the North American Chapter of the Association for Computational Linguistics: Human Language Technologies, pp. 1112–1122. Association for Computational Linguistics (2018)

18. Rajpurkar, P., Zhang, J., Lopyrev, K., Liang, P.: Squad: 100, 000+ questions for machine comprehension of text. In: Proceedings of the 2016 Conference on Empirical Methods in Natural Language Processing, pp. 2383–2392. Association for Computational Linguistics (2016)

19. He, L., Lewis, M., Zettlemoyer, L.: Question-answer driven semantic role labeling: using natural language to annotate natural language. In: Proceedings of the 2015 Conference on Empirical Methods in Natural Language Processing, pp. 643–653. Association for Computational Linguistics (2015)
20. Socher, R.: Recursive deep models for semantic compositionality over a sentiment treebank. In: Proceedings of the 2013 Conference on Empirical Methods in Natural Language Processing, pp. 1631–1642. Association for Computational Linguistics (2013)
21. Zhong, V., Xiong, C., Socher, R.: Seq2SQL: generating structured queries from natural language using reinforcement learning. CoRR, vol. abs/1709.00103 (2017)
22. Mrksic, N., Vulic, I.: Fully statistical neural belief tracking. In: Proceedings of the 56th Annual Meeting of the Association for Computational Linguistics, pp. 108–113 (2018)
23. Levy, O., Seo, M., Choi, E., Zettlemoyer, L.: Zero-shot relation extraction via reading comprehension. In: Proceedings of the 21st Conference on Computational Natural Language Learning (CoNNL 2017), pp. 333–342 (2017)
24. Lin, C.-Y.: Rouge: a package for automatic evaluation of summaries. In: Text Summarization Branches Out: Proceedings of the ACL-2004 Workshop, vol. 8, pp. 74–81 (2004)
25. Chen, V.: An examination of the CNN/DailyMail neural summarization task. In: Proceedings of the 54th Annual Meeting of the Association for Computational Linguistics, pp. 2358–2367 (2017)
26. Rush, A.M., Chopra, S., Weston, J.: A neural attention model for abstractive sentence summarization. In: Proceedings of the 2015 Conference on Empirical Methods in Natural Language Processing, pp. 379–389 (2015)
27. Chopra, S., Auli, M., Rush, A.M.: Abstractive sentence summarization with attentive recurrent neural networks. In: Proceedings of the 2016 Conference of the North American Chapter of the Association for Computational Linguistics: Human Language Technologies, pp. 93–98. Association for Computational Linguistics (2016)

Highlighted Word Encoding
for Abstractive Text Summarization

Daisy Monika Lal[✉], Krishna Pratap Singh, and Uma Shanker Tiwary

Machine Learning and Optimization Lab, Deptartment of Information Technology,
IIIT Allahabad, Prayagraj, India
daisylal26@gmail.com, {kpsingh,ust}@iiita.ac.in

Abstract. The proposed model unites the robustness of the extractive and abstractive summarization strategies. Three tasks indispensable to automatic summarization, namely, apprehension, extraction, and abstraction, are performed by two specially designed networks, the highlighter RNN and the generator RNN. While the highlighter RNN collectively performs the task of highlighting and extraction for identifying the salient facts in the input text, the generator RNN fabricates the summary based on those facts. The summary is generated using word-level extraction with the help of term-frequency inverse document frequency (TFIDF) ranking factor. The union of the two strategies proves to surpass the ROUGE score results on the Gigaword dataset as compared to the simple abstractive approach for summarization.

Keywords: Extractive summarization · Abstractive summarization · Sequence-to-sequence modeling · Attention mechanism · Tf_idf score

1 Introduction

Automatic Text Summarization is a requisite for natural language understanding. Summaries relay all the important information in a time-saving manner. This makes summarization an extremely essential task specifically to deal with the fortune of information available online. Summarization involves mapping the input to purposeful interpretations, inspecting different aspects of the input and processing it to generate meaningful phrases and sentences. Summaries can be generated using *Extractive* or *Abstractive* approaches. Extractive methods bring into the spotlight phrases or sentences that display the most valuable pieces of information for a given input. The output is an amalgamation of the identified keywords, phrases and sentences. The abstractive method tries to gain insights about the input text and concludes with a shortened version of it restated using words of its own while preserving the authenticity of the information contained in the original text. With respect to the *ROUGE* estimate, extractive techniques have surpassed the abstractive ones. But extraction and abstraction are both crucial to the task of summarization. While extraction ensures bringing forth the most relevant portions of text, abstraction makes sure all those portions are fused together into a meaningful abstract.

© Springer Nature Switzerland AG 2020
U. S. Tiwary and S. Chaudhury (Eds.): IHCI 2019, LNCS 11886, pp. 77–86, 2020.
https://doi.org/10.1007/978-3-030-44689-5_7

Our model is the unison of both strategies. It is composed of two recurrent neural networks, the Extractor RNN, and the Abstractor RNN. The extractor RNN, which is a single-layer RNN, calculates the value scores for each word and based on a threshold score outputs the encoded representations of the appropriate words. The abstractor RNN outputs the summary, word by word, using the importance score and *Attention Mechanism* [1].

2   Related Work

Extractive Methods generate summaries based on value scores. A subset of words is selected from the input text and directly copied into the output summary. Extractive summarization algorithms, in the past, placed their focus on unsupervised approaches that fall into three broad categories: *statistical* [2], *graph-based* [3], and *optimization-based approaches* [4]. Recent advances in deep learning have led to supervised models such as Neural Networks being adopted for the task of summarization. Supervised extractive summarization algorithms treat the task of summarization as a classification problem. The main objective is to decide whether or not a word, phrase or sentence should be included in the summary. *Cheng et al.* [6] suggested a data-driven approach that extracts sentences using sentence-level encodings and an attention-based extractor. *Nallapati et al.* [7] proposed two separate architectures *Classify*: This approach involves reading the input once, for gaining an understanding of the text. The input is traversed once again, picking out sentences for the summary, in the order of traversal. *Select*: This approach involves traversing through the input and memorizing the significant information. The summary is then generated from scratch by picking out sentences at random. *Narayan et al.* [8] recommended a model that summarizes based on side-information. The authors lay emphasis on the fact that side-information like images, titles, image captions, etc. can help in producing a more informative summary. To date, the *Summarunner* introduced by *Nallapati et al.* [5] outperforms the other extraction-based summarizers.

Abstractive Text Summarization attempts to sum up the crux of a given text in limited words. It relies on its memory to recall all the important details while generating the summary. Variants of Encoder-Decoder models have been recycled to create abstraction-based summaries. *Rush et al.* [11] suggested an Attention-Based Summarization (ABS) to output a summary where each word of the summary is conditioned on the input sequence. *Nallapati et al.* [9] experimented with the basic Encoder Decoder Attention Model in four different scenarios: (a) *Large Vocabulary Trick (LVT)* that reduces the size of the decoder-vocabulary to achieve computational efficiency, (b) *Feature-Rich Encoded Representations* to capture the essence of the source more accurately, (c) *Handling Rare and Unseen Word Problem* using pointer-networks which copies words directly from the input, (d) *Hierarchical Attention* to create sentence-level encoded representations. *Paulus et al.* [10] proposed a model for long-document summarization using intra-temporal attention. The model, at each decoding step, takes into consideration not only the set of encoder hidden states but also the decoder hidden states. The model is specifically designed to address the problem of repetitive phrases and sentences. ***Zhou et al.*** [13] recommended a two-level sentence encoding technique that makes use of a selective gate network to construct second-level sentence representations from the first-level sentence representations. ***Fan et al.*** [12] recently introduced the concept of controllable summarization that uses a Convolutional Neural Network for encoding and decoding. It facilitates length-constrained, entity-centric and source-specific summarization.

3   Background

The three inescapable steps [Fig. 1] involved in the task of text summarization include:

1. Recognizing the Salient Information (***Highlighting***).
2. Recording the Salient Information (***Extraction***).
3. Writing from the Memory (***Abstraction***).

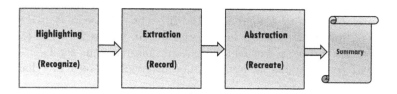

Fig. 1. Summarization steps

3.1   Highlighting: Recognizing the Salient Information

Highlighting is the fundamental step towards summarizing any piece of literature. It requires one to be thorough with the piece. The purpose is to read and identify the most informative bits in the given text. The task of recognizing the crucial details requires that the text be read with meticulous care.

3.2   Extraction: Recording the Salient Information

The second step is to outline the work. For this, the main ideas presented in the piece need to be identified and recorded. The process of identifying what is of essential importance for defining a piece of text is called Extraction.

3.3   Abstraction: Writing from the Memory

Once the main ideas in the text have been discovered, the next and the final step is to start expanding on those ideas. This step requires recalling from memory in order to avoid the repetition of words and phrases present in the original text and also for obtaining clarity. The process of unfolding on the main ideas or revealing relevant facts for a given piece of text is called Abstraction. Since all the three steps are mandatory for writing a comprehensive summary, there is a need for collaborating extractive techniques with the abstractive ones.

4   Model: *Word-Level Highlighter-Generator Network*

Our model unifies extraction and abstraction mechanisms in order to generate a summary. The three steps are realized using two recurrent neural networks: *Highlighter and Generator*.

Highlighter Network: The Highlighter RNN reads the input text X and identifies the important words based on an importance score (tf_idf value). These identified fragments are saved as highlighted text called the **Extract**. The extract is expected to encompass all the relevant information pertaining to the work. The importance score of each fragment is calculated based on the **Term-Frequency and Inverse Document Frequency (tf_idf)**. The tf_idf score is used to identify the most informative terms in a document. It is based on the fact that infrequent terms add more factual information than recurring terms.

The **term-frequency** is the number of occurrences of a term in a document. It is calculated as:

$$tf_t = \frac{n_t}{N} \tag{1}$$

where, n_t = number of times a term t appears in a sentence and N = total number of terms in the sentence.

The **inverse document frequency** is a measure of the rarity of a term in a document. It is calculated as:

$$idf_t = log_e \frac{S}{s_t} \tag{2}$$

where, S = total number of sentences and s_t = number of sentences with term t in it. Both the scores are calculated on the **log-scale** as:

$$tf_t = 1 + log_e(1 + \frac{n_t}{N}) \tag{3}$$

$$idf_t = log_e \frac{S}{s_t} \tag{4}$$

The **tf_idf** score is calculated as:

$$(tf\_idf)_t = tf_t * idf_t \tag{5}$$

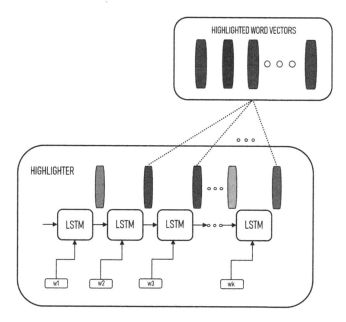

Fig. 2. Word-Level Highlighter

The *highlighter* [Fig. 2] is an *LSTM* cell that takes as input the source article, word by word, $X = (x_1, x_2, ..., x_{Tx})$ where Tx is the length of the source article and generates the hidden representation h_t for each word [Eq. 6]. It then uses the previously calculated $tf\_idf\ scores$ for each word to determine which encoded hidden representations should be selected for the decoding process. The highlighter being an LSTM network, is capable of capturing long-term dependencies. This in turn, enhances the efficiency of the model. Different variants of recurrent neural networks like the Gated Recurrent Unit (GRU), Bidirectional LSTMs and GRUs can also be used as the highlighter network.

$$f_t = \sigma(W_f[h_{t-1}, x_t] + b_f)$$
$$i_t = \sigma(W_i[h_{t-1}, x_t] + b_i)$$
$$\tilde{c}_t = tanh(W_c[h_{t-1}, x_t] + b_c)$$
$$c_t = (f_t \times c_{t-1}) + (i_t \times \tilde{c}_t) \qquad (6)$$
$$o_t = \sigma(W_o[h_{t-1}, x_t] + b_o)$$
$$h_t = o_t \times tanh(c_t)$$

Generator Network: The generator RNN begins to write the abstract from the memory (extracted words). It aligns the set of highlighted words with each output word y_t. The entire text is not considered while generating the summary, instead, only the extracted words. This ensures that the summary is not only concise but also captures all the crucial information that must be conveyed to the reader. The selected hidden vectors $(high_1, high_2, ..., high_k)$, where k is the number of selected hidden states, are sent to the generator for summary generation. The *generator* takes as input the highlighted vectors and the the actual output summary, word by word, to produce the summary using attention mechanism [Eq. 7].

$$\tilde{h}_t = tanh(W_a[\tilde{h}_{t-1}, y_t] + b_a)$$

$$\alpha_{ti} = \frac{exp(s_{ti})}{\sum_{j=1}^{k} exp(s_{tj})}$$

$$s(high_i, \tilde{h}_t) = (high_i)^T (\tilde{h}_t) \tag{7}$$

$$\tilde{z}_t = \sum_{j=1}^{k} \alpha_{tj}(high_j)$$

$$C_t = W_{\tilde{z}}(\tilde{z}_t \bigoplus \tilde{h}_t) + b_{\tilde{z}}$$

4.1   Algorithm for Highlighting and Extracting Terms

The following steps are involved in the process of extraction at the word-level:

Algorithm 1: *Word-Level Highlighter*

Result: *Highlighted Hidden State Vectors*
init_highlighter_hidden_state = **0**;
$H_{thres} = w$;
while X_i **do**
  $\quad$ *word_tokenize*;
  $\quad$ **while** x_{ij} **do**
  $\quad\quad$ | *compute tf_idf$_{ij}$*;
  $\quad$ **end**
  $\quad$ **while** x_{ij} **do**
  $\quad\quad$ *read Ex$_{ij}$*;
  $\quad\quad$ *compute hidden_state_vectors* h_{ij}^{high};
  $\quad\quad$ **if** *tf_idf$_{ij} \geq H_{thres}$* **then**
  $\quad\quad\quad$ | *save hidden_state_vector h_{ij}^{high} in the highlighter network*;
  $\quad\quad$ **else**
  $\quad\quad\quad$ | *discard h_{ij}^{high}*;
  $\quad\quad$ **end**
  $\quad$ **end**
end

---

where, X_i : i^{th} *input training example,* x_{ij} : j^{th} *word of the ith input training example,* Ex_{ij} : *Embedding of the jth word of the ith input training example,* *tf_idf$_{ij}$* : *term-frequency inverse document frequency of the jth word of the ith input training example,* h_{ij}^{high} : *hidden_state_vector of the jth word of the ith input training example,* H_{thres} : *threshold value for highlighting a word,*

5   Experiments and Results

5.1   Dataset

The summarizer was trained on the **Gigaword** dataset which was first used by **Rush et al.** [11]. It comprises of news-headline pairs collected from several different news sources including the **New York Times Newswire Service** and **Los Angeles Times/Washington Post Newswire Service**. There are about **3.8M** training, **189k** development, and **1951** test instances of news articles available in the dataset which were sampled randomly for training the model.

5.2   Experimental Setup

□ The proposed method is implemented in **Python 3.6**. The following python libraries were used:

  ⋆ **Keras**
  ⋆ **Pandas**
  ⋆ **Numpy**
  ⋆ **Nltk**

□ The system configurations for the experiment were as follows:

  ⋆ **Intel Xeon Fourth Generation Processor**
  ⋆ **12 GB Nvidia Geforce GTX 1080ti**
  ⋆ **Windows 10 pro**
  ⋆ **128 GB RAM**

5.3   Evaluation Metric

ROUGE [14], an acronym for **Recall-Oriented Understudy for Gisting Evaluation**, is a package used to evaluate the quality of the machine-generated summaries against a human-generated summary. This evaluation metric uses **Recall** and **Precision** to determine whether or not a machine-generated summary is accurate and reliable. **Recall** measures the number of words common to both the machine-generated summary and human-generated summary given the total number of words in the human-generated summary.

Let, **Overlap**: *number of overlapping words,*
N_h : *total number of words in the human-generated summary* and
N_m: *total number of words in the machine-generated summary.*

$$Recall = \frac{Overlap}{N_h} \qquad (8)$$

$$Precision = \frac{Overlap}{N_m} \qquad (9)$$

The harmonic mean of both the Recall and Precision is used to finally determine the quality of the automatically generated summary. This is called the
F-Measure.

$$F - Measure = \frac{2 \times Recall \times Precision}{Recall + Precision} \qquad (10)$$

Our model is assessed on three variants of ROUGE, namely, **Rouge-1:** Unigram recall between the actual summary and the model generated summary,
Rouge-2: Bi-gram recall between the actual summary and the model generated
summary, **Rouge-L:** Longest Common Subsequence (LCS) based F-Measure
between the actual summary and the model generated summary.

5.4   Results

In this setup a batch size of 150, embedding size of 300 and 1000 hidden neurons
were chosen for training. The model takes around 12 min per 1000 epochs. For
better convergence and accuracy the model was trained for 200000 epochs and
validated with 10000 articles using **Glove** pre-trained embedding.

The performance comparison of our model with various other models is presented in Table [1]. On the Gigaword corpus our model performs slightly better
than the ABS and ABS+ models presented by **Rush et al.** [11] and the RAS-
Elman and RAS-LSTM models presented by **Chopra et al.** [15].

Table 1. Results on Gigaword test set using F1 ROUGE metrics. k refers to the
beam search size.

MODEL	Rouge-1	Rouge-2	Rouge-L
ABS [11]	29.55	11.32	26.42
ABS+ [11]	29.76	11.88	26.96
RAS-Elman (k = 1) [15]	33.10	14.45	30.25
RAS-Elman (k = 10) [15]	33.78	15.97	31.15
RAS-LSTM (k = 1) [15]	31.71	13.63	29.31
RAS-LSTM (k = 10) [15]	32.55	14.70	30.03
Our Model	**34.77**	**24.16**	**28.82**

Below is a comparison between the actual summaries and the summaries generated by our model:

Reference Article 1:
"At least two people were killed in a suspected bomb attack on a passenger bus in the strife-torn southern philippines"
Actual Summary:

"At least two dead in southern philippines blast"
Generated Summary:

"two killed in bomb attack in southern philippines"

Reference Article 2:

"spanish property group colonial , struggling under huge debts , announced losses of . billion euros for the first half of which it blamed on asset depreciation ."
Actual Summary:

"spain 's colonial posts #.## billion euro loss"
Generated Summary:

"spain 's colonial posts #.## billion billion billion"

Reference Article 3:
"south korea's nuclear envoy Kim sook urged north korea monday to restart work to disable its nuclear plants and stop its " typical " brinkmanship in negotiations ."
Actual Summary:

"envoy urges north korea to restart nuclear disablement"
Generated Summary:

"south korea's nuclear envoy urges north korea to restart work"

6   Conclusion and Future Work

The unified model along with GloVe Embedding has proved to generate better results for the Gigaword Datatset than the simple abstractive text summarization model. The input texts in this Gigaword dataset comprise of only a few sentences and since the tf_idf score is more effective when the number of input sentences is large, the unified model is yet to be tested on datasets with large input length or multiple paragraphs like the CNN/Daily Mail. In addition, since the model is applied at the word-level it would also be interesting to see the results when it is applied at the sentence-level.

References

1. Bahdanau, D., Cho, K., Bengio, Y.: Neural machine translation by jointly learning to align and translate. CoRR, abs/1409.0473 (2014)
2. Saggion, H., Poibeau, T.: Automatic Text Summarization: Past Present and Future, pp. 3–13. Springer, Heidelberg (2016)
3. Barrios, F., Lopez, F., Argerich, L., Wachenchauzer, R.: Variations of the Similarity Function of TextRank for Automated Summarization. arXiv preprint arXiv:1602.03606v1 (2016)
4. Durrett, G., Berg-Kirkpatrick, T., Klein, D.: Learning-Based Single-Document Summarization with Compression and Anaphoricity Constraints. arXiv preprint arXiv:1603.08887v2 (2016)
5. Nallapati, R., Zhai, F., Zhou, B.: Summarunner: a recurrent neural network-based sequence model for extractive summarization of documents (2017)
6. Cheng, J., Lapata, M.: Neural summarization by extracting sentences and words. In: Proceedings of the 54th Annual Meeting of the Association for Computational Linguistics (Volume 1: Long Papers), vol. 1, pp. 484–494 (2016)
7. Nallapati, R., Zhou, B., Ma, M.: Classify or select: Neural architectures for extractive document summarization. arXiv preprint arXiv:1611.04244 (2016a)
8. Narayan, S., Papasarantopoulos, N., Lapata, M., Cohen, S.B.: Neural extractive summarization with side information. arXiv preprint arXiv:1704.04530 (2017)
9. Nallapati, R., Zhou, B., dos Santos, C., Gulcehre, C., Xiang, B.: Abstractive text summarization using sequence-to-sequence RNNs and beyond. In Proceedings of the 20th SIGNLL Conference on Computational Natural Language Learning, pp. 280–290 (2016b)
10. Paulus, R., Xiong, C., Socher, R.: A deep reinforced model for abstractive summarization. In: Proceedings of the 2018 International Conference on Learning Representations (2017)
11. Rush, A.M., Chopra, S., Weston, J.: A neural attention model for abstractive sentence summarization. In: Proceedings of the 2015 Conference on Empirical Methods in Natural Language Processing, pp. 379–389 (2015)
12. Fan, A., Grangier, D., Auli, M.: Controllable abstractive summarization. arXiv preprint arXiv:1711.05217 (2017)
13. Zhou, Q., Yang, N., Wei, F., Zhou, M.: Selective Encoding for Abstractive Sentence Summarization. arXiv preprint arXiv:1704.07073v1 (2017)
14. Lin, C.-Y.: ROUGE: a package for automatic evaluation of summaries. In: Proceedings of the ACL Workshop: Text Summarization Branches Out (2004)
15. Chopra, S., Auli, M., Rush, A.M.: Abstractive sentence summarization with attentive recurrent neural networks. In: Proceedings of the ACL Workshop: Text Summarization Branches Out (2016)

Detection of Hate and Offensive Speech in Text

Abid Hussain Wani$^{(\boxtimes)}$ (iD), Nahida Shafi Molvi, and Sheikh Ishrah Ashraf

Department of Computer Science, University of Kashmir, South Campus, Srinagar, India
abid.wani@uok.edu.in, nahidashafi@mail.com,
sheikhishrah@gmail.com

Abstract. With online social platforms becoming more and more accessible to the common masses, the volume of public utterances on a range of issues, events, and persons etc. has increased profoundly. Though most of the content is a manifestation of personal feelings of the individuals, yet a lot of this content often comprises of hate and offensive speech. Exchange of hate and offensive speech has now become a global phenomenon with increased intolerance among societies. However companies running these social media platforms need to discern and remove such unwanted content. This article focuses on automatic detection of hate and offensive speech from Twitter data by employing both conventional machine learning algorithms as well as deep learning architectures. We conducted extensive experiments on a benchmark 25K Twitter dataset with traditional machine learning algorithms as well as using deep learning architectures. The results obtained by us using deep learning architectures are better than state-of-the-art methods used for hate and offensive speech detection.

Keywords: Text classification · Offensive text · Deep learning

1 Introduction

Ease of access to social media platforms propelled by ubiquitous use of Internet-enabled mobile devices paves way for the exchange of positive feelings and ideas on the one side but on the other side it is subject to misuse and hate-mongering as well. Online social media platforms provide an easy and global means to express oneself be it normal online discourse or hate speech. Hate Speech comprises of abusive utterances usually targeted towards an individual, a group, an organization etc. to insult, incite hatred, or cause damage to reputation.

An important consideration for today's online social platforms is the automatic detection of hate and offensive utterances in natural languages whose nature can be quite weird and often difficult to detect. Although a number of popular social networking giants like Twitter and Facebook have engaged human workforce to review online content so as to remove abusive and offensive content, yet the volume of content to be reviewed is so huge that it is humanly impossible to review it on the fly. Therefore there is great scope for enhancing automatic hate and offensive detection systems.

The detection of hate-filled and abusive language in text is modelled as a classification task and has been approached through a number of techniques and algorithms. In this

© Springer Nature Switzerland AG 2020
U. S. Tiwary and S. Chaudhury (Eds.): IHCI 2019, LNCS 11886, pp. 87–93, 2020.
https://doi.org/10.1007/978-3-030-44689-5_8

work, we carried out a number of experiments on a standard dataset of 25K tweets [1]. In our knowledge, combinations of the set of features used and the techniques employed by us make our work distinct from other similar studies. We experimented with conventional machine learning classifiers like Decision Tree, Support Vector Machine etc. as well as deep learning architectures like Convolutional Neural Networks and Long Short-Term Memory Networks. The classification performance achieved by us using deep learning architectures is better than previous similar studies.

2   Related Work

Given the scope of its applications, the task of hate speech detection has received much attention from researchers. In our knowledge, much of the work in this direction has been carried out with an emphasis on using language features together with machine learning techniques. Studies employing only lexical features fail to capture the context of the sentences. N-gram based and semantic-relatedness based approaches, on the otherhand have been found to perform better [2, 3]. Like in most text classification tasks, most of the work in hate speech detection is based on supervised learning. Support Vector Machine, Logistic Regression etc. have been the predominant classifiers employed by a number of studies [4, 5]

Deep learning-based methods encompass learning abstract feature representation from text corpus using multiple stacked neural layers. An interesting work in this direction uses both Convolutional Neural Network as well as Long Short Term Memory (Recurrent Neural Network) [6, 12].

3   Experiments

For the detection of hate-filled tweets, we conducted experiments employing conventional machine learning techniques as well as state-of-the-art deep learning architectures. All our experiments were performed on a 25K Twitter dataset with three labels: HATE, OFFENSIVE and OK [1]. Tweets containing hateful content are annotated with HATE label, those encompassing only offensive content are annotated with OFFENSIVE label and the tweets that include neither hateful nor offensive text utterances have been labelled as OK in this dataset. In the following subsections, we discuss the experimental methodology employed for both conventional machine and deep learning architectures.

3.1   Detecting Hate Using Conventional Machine Learning Methods

In order to carry out various machine learning tasks on the data, we first put it to preprocessing to get rid of all unwanted stuff from the text corpus. The basic text preprocessing steps are undertaken on each tweet to remove the unnecessary contents from the tweets including URLs, Email IDs, Punctuations, Stopwords, Space Patterns. To get rid of word-suffixes and inflexional endings we employ Porter Stemmer [8]. Traditional machine learning methods work by first extracting features from the textual data instances before attempting to classify them. In this work we employed the following feature groups for classification by four conventional classifiers as enumerated below:

- Logistic Regression with Char4-grams
- Support Vector Machine with Bi-grams, BagofWordsVectorizer and TFIDF
- Naive Bayes Classifier with Bi-grams, BagofWordsVectorizer and TFIDF
- Decision Tree Classifier with Bi-grams, BagofWordsVectorizer and TFIDF

Studies on classification tasks have shown that simple text features such as character n-grams, word n-grams, and bag of words are very effective in predicting the actual label of a textual data instance [9, 10].

To compare the classification performance of the above classic classifiers, we computed the metrics of Precision, Recall and F1-score for each of the classifiers as depicted in Table 1. Using the features listed above, we achieved the best results by employing Decision Tree Classifier and Support Vector Classifier. The confusion matrices obtained on the dataset for Decision Tree Classifier and Support Vector Machine Classifier are shown in Figs. 1 and 2 respectively.

Apart from surface textual features as noted above, we conducted experiments employing pre-trained GloVe (Global Vectors for Word Representation) [7] as well as Word2Vec word embeddings[1].

Table 1. Classification results using conventional machine learning methods

Method	Precision	Recall	F-Score
LogisticRegressionClassifier + Char4-grams	0.729	0.778	0.752
LogisticRegressionClassifier + BagofWordsVectorizer + TFIDF	0.812	0.801	0.806
SupportVectorClassifier + bi-grams + BagofWordsVectorizer + TFIDF	0.812	0.801	0.806
NaiveBayesClassifier + bi-grams + BagofWordsVectorizer + TFIDF	0.791	0.810	0.800
DecisionTreeClassifier + bi-grams + BagofWordsVectorizer + TFIDF	0.820	0.821	0.820

3.2 Detecting Hate and Offensive Speech Using Deep Learning

Deep Learning architectures have recently gained much attention in a number of areas and text classification is no exception. After undertaking pre-processing of text corpora as done in case of conventional methods. To detect hate and offensive speech in social media text (Twitter data in our experiments), we experiment with two neural network architectures. The word embeddings employed in our experiments include (Fig. 3):

- Random Embeddings
- GloVe Embeddings
- Word2Vec Embeddings

---

[1] https://github.com/mmihaltz/word2vec-GoogleNews-vectors.

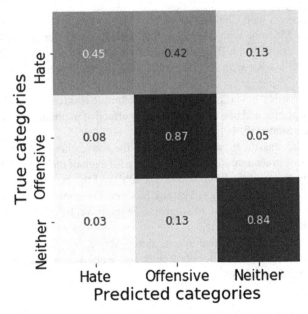

Fig. 1. Confusion matrix obtained for decision tree hate speech classifier.

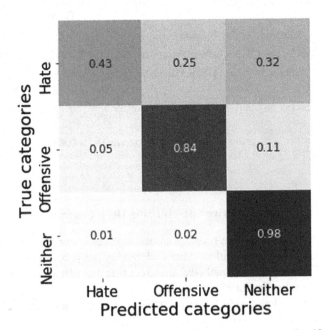

Fig. 2. Confusion matrix obtained on applying support vector classifier.

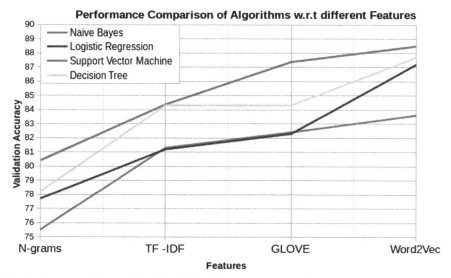

Fig. 3. Accuracy of conventional classifiers in hate speech detection for different feature groups.

We employ the same settings for CNN as used in [11] for sentence classification. In our baseline model, we employ CNN-rand variant of CNN wherein all words are randomly initialized and then modified during training. In our experiment we employ: rectified linear units, filter windows (h) of 3, 4, 5 with 100 feature maps each, dropout rate (p) of 0.5 and mini-batch size of 50. In addition to CNN, our hate and offensive speech classification framework encompasses LSTM. LSTMs are "recurrent neural networks that use their internal memory to process arbitrary sequences of inputs". In effect, with LSTMs we take care of "long range dependencies in text" to improve the classification process. Table 2 depicts the classification performance metrics obtained using different word embeddings. Figure 4 compares the performance of conventional machine learning to that of deep learning approaches for the detection of hate speech in text.

Table 2. Comparison of various deep learning methods (embedding size = 200)

Method	Precision	Recall	Fscore	Accuracy
ConvolutionalNeuralNetwork + Glove	0.812	0.802	0.801	82.20%
ConvolutionalNeuralNetwork + Word2vec	0.870	0.850	0.910	92.37%
ConvolutionalNeuralNetwork + RandomEmbeddings	0.822	0.821	0.831	92.01%
LongShortTermMemory + Glove	0.850	0.8508	0.861	93.01%
LongShortTermMemory + Word2vec	0.920	0.920	0.915	93.75%
LongShortTermMemory + RandomEmbeddings	0.920	0.920	0.915	93.45%

RESULTS AND ANALYSIS

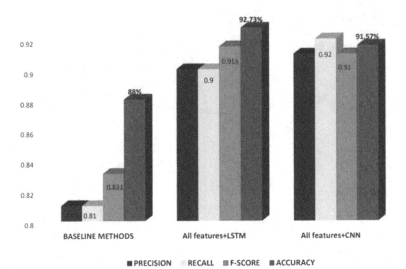

Fig. 4. Performance comparison of machine learning and deep learning approaches to hate and offensive speech detection.

4   Conclusion

The importance of automatic hate and offensive speech detection systems has ever increased in the recent times with global reach of social networking sites and applications. Our work made several contributions in this area. First, we presented the results of hate and offensive speech detection using conventional machine learning techniques using various feature groups. We found that the Decision-Tree-Classifier using bi-grams with Bag-of-Words Vectorizer and TFIDF resulted in highest F1 score and for other classifiers the results were almost similar except for Logistic Regression Classifier with Char4-grams. Second, we conducted several experiments using a deep neural network architectures namely CNN and LSTM to improve classification accuracy employing three different word embeddings. Our experiments show that for hate speech detection Long Short Term Memory with Word2vec provides the highest classification performance. Finally, we presented a comparative performance evaluation of baseline classic machine learning methods and deep learning methods for the task of hate and offensive speech detection.

References

1. Davidson, T., Warmsley, D., Macy, M., Weber, I.: Automated hate speech detection and the problem of offensive language. In: 11th International AAAI Conference on Web and Social Media., Montreal (2017)
2. Nobata, C., Tetreault, J., Thomas, A., Mehdad, Y., Chang, Y.: Abusive language detection in online user content. In: Proceedings of the 25th International Conference on World Wide Web - WWW 2016, Montral, Qubec, Canada (2016)

3. Silva, L.A., Mondal, M., Correa, D., Benevenuto, F., Weber, I.: Analyzing the targets of hate in online social media. In: ICWSM, pp. 687–690 (2016)

4. Mehdad, Y., Tetreault, J.: Do characters abuse more than words? In: Proceedings of the SIGDIAL 2016 Conference, Los Angeles, USA, pp. 299–303. Association for Computational Linguistics (2016)

5. Waseem, Z.: Are you a racist or am i seeing things? Annotator influence on hate speech detection on Twitter. In: Proceedings of the First Workshop on NLP and Computational Social Science, Austin, Texas, pp. 138–142. Association for Computational Linguistics (2016)

6. Wei, X., Lin, H., Yang, L., Yu, Y.: A convolution-LSTM-based deep neural network for cross-domain MOOC forum post classification. Information **8**(3), 92 (2017)

7. Pennington, J., Socher, R., Manning, C.D.: GloVe: global vectors for word representation. In: EMNLP, vol. 14, pp. 1532–1543 (2014)

8. Porter, M.F.: An algorithm for suffix stripping. Program **14**, 130–137 (1980)

9. Greevy, E., Smeaton, A.F.: Classifying racist texts using a support vector machine. In: Proceedings of the 27th Annual International ACM SIGIR Conference on Research and Development in Information Retrieval (SIGIR 2004), pp. pp. 468–469. ACM, New York (2004). https://doi.org/10.1145/1008992.1009074

10. Del Vigna, F., Cimino, A., Orletta, F.D., Petrocchi, M., Tesconi, M.: Hate me, hate me not: hate speech detection on Facebook. In: Proceedings of the First Italian Conference on Cyber security, Venice, Italy, pp. 86–95 (2017)

11. Kim, Y.: Convolutional neural networks for sentence classification. In: EMNLP, pp. 1746–1751 (2014)

12. Fortuna, P., Nunes, S.: A survey on automatic detection of hate speech in text. ACM Comput. Surv. (CSUR) **51**(4), 85 (2018)

Deep Learning for Hindi Text Classification: A Comparison

Ramchandra Joshi[1], Purvi Goel[2], and Raviraj Joshi[2(✉)]

[1] Department of Computer Engineering,
Pune Institute of Computer Technology, Pune, India
`rbjoshi1309@gmail.com`
[2] Department of Computer Science and Engineering,
Indian Institute of Technology Madras, Chennai, India
`goyalpoorvi@gmail.com`, `ravirajjoshi@gmail.com`

Abstract. Natural Language Processing (NLP) and especially natural language text analysis have seen great advances in recent times. Usage of deep learning in text processing has revolutionized the techniques for text processing and achieved remarkable results. Different deep learning architectures like CNN, LSTM, and very recent Transformer have been used to achieve state of the art results variety on NLP tasks. In this work, we survey a host of deep learning architectures for text classification tasks. The work is specifically concerned with the classification of Hindi text. The research in the classification of morphologically rich and low resource Hindi language written in Devanagari script has been limited due to the absence of large labeled corpus. In this work, we used translated versions of English data-sets to evaluate models based on CNN, LSTM and Attention. Multilingual pre-trained sentence embeddings based on BERT and LASER are also compared to evaluate their effectiveness for the Hindi language. The paper also serves as a tutorial for popular text classification techniques.

Keywords: Natural language processing · Convolutional neural networks · Recurrent neural networks · Sentence embedding · Hindi text classification

1 Introduction

Natural language processing represents computational techniques used for processing human language. The language can either be represented in terms of text or speech. NLP in the context of deep learning has become very popular because of its ability to handle text which is far from being grammatically correct. Ability to learn from the data have made the machine learning system powerful enough to process any type of unstructured text. Machine learning approaches have been used to achieve state of the art results on NLP tasks like text classification, machine translation, question answering, text summarization, text ranking, relation classification, and others.

© Springer Nature Switzerland AG 2020
U. S. Tiwary and S. Chaudhury (Eds.): IHCI 2019, LNCS 11886, pp. 94–101, 2020.
https://doi.org/10.1007/978-3-030-44689-5_9

The focus of our work is text classification of Hindi language. Text classification is the most widely used NLP task. It finds application in sentiment analysis, spam detection, email classification, and document classification to name a few. It is an integral component of conversational systems for intent detection. There have been very few text classification works in literature focusing on the resource-constrained Hindi language. While the most important reason for this is unavailability of large training data; another reason is generalizability of deep learning architectures to different languages. However, Hindi is morphologically rich and relatively free word order language so we investigate the performance of different models on Hindi text classification task. Moreover, there has been a substantial rise in Hindi language digital content in recent years. Service providers, e-commerce industries are now targeting local languages to improve their visibility. Increase in the robustness of translation and transliteration systems have also contributed to the rise of NLP systems for Hindi text. This work will help in the selection of right models and provide a suitable benchmark for further research in Hindi text classification tasks.

In order to create Hindi dataset in Devanagari script, standard datasets like TREC, SST were translated using Google translate. This is indeed great times for translation systems. Self attention-based models like transformer have resulted in best in class results for translation tasks. We believe this is the right time to evaluate the translated data sets. Even the multi-lingual datasets like XNLI [3] used for evaluation of natural language inference tasks is based on the translation.

Current text classification algorithms are mainly based on CNNs and RNNs. They work at the sentence level, paragraph level or document level. In this work, we consider sentence-level classification tasks. Each sentence split into a sequence of word tokens and passed to classification algorithms. Instead of passing the raw character tokens, each word is mapped to a numerical vector. The sequence of vectors is processed by classification algorithms. A very common approach is to learn these distributed vectorial representations using unsupervised learning techniques like word2vec [8]. The similarity of the word vectors are correlated to the semantic similarity between actual words. This gives some useful semantic properties to low dimensional word vectors. Usage of pre-trained vectors have shown to give superior results and are thus the de-facto method to represent word tokens in all NLP models. In this work, we use FastText word vectors, pre-trained on Hindi corpus. The embedding matrix is used as an input to deep learning models. Naive bag of words approach is to average the word embeddings and then use a linear classifier or feed-forward neural network to classify the resulting sentence embedding. A more sophisticated approach is to pass the sequence of word vectors through LSTM and use final hidden state representation for classification. CNNs are also pretty popular for sentence classification tasks where a fixed length padded word vector sequence is passed through the CNNs. We explore different variations of LSTM, CNN, and Attention-based neural networks for comparison.

Learning universal sentence representations is another area of active research. These sentence representations are used for classification tasks. General idea is to use a large amount of labeled or unlabelled corpus to learn sentence representations in a supervised or unsupervised setting. This is similar to learning word vectors externally and using them in the target task. These approaches represent transfer learning in the context of NLP. Models like Skip-Thought Vectors, Universal Sentence Encoder by Google, InferSent, and BERT have to be used to learn sentence embeddings. Using pre-trained sentence embeddings lowers the training time and is more robust on small target data sets. In this work, we also evaluate pre-trained multi-lingual sentence embedding obtained using BERT and LASER to draw a better comparison.

Main contributions of this paper are:

- Compare variations of CNN and LSTM models for Hindi text classification.
- Effectiveness of Hindi Fast-Text word embedding is evaluated.
- Effectiveness of multi-lingual pre-trained sentence embedding based on BERT and LASER is evaluated on Hindi corpus.

2   Related Work

There has been limited literature on Hindi text classification. Arora Piyush in his early work [1] used traditional n-gram and weighted n-grams method for sentiment analysis of Hindi text. Tummalapalli *et al.* [9] used deep learning techniques- basic CNN, LSTM and multi-Input CNN for evaluating the classification accuracy of Hindi and Telugu texts. Their main focus was capturing morphological variations in Hindi language using word-level and character-level features. CNN based models performed better as compared to LSTM and SVM using n-gram features. The datasets used were created using translation. In this work, we are concerned with the performance of different model architectures and word vectors so we do not consider character level or subword level features.

In general, there has been a lot of research on text classification and sentiment analysis employing supervised and sem-supervised techniques. Kim *et al.* [6] proposed CNN based architecture for classification of English sentences. A simple bag of words model based on averaging of fast text word vectors was proposed in [5]. They proposed a simple fast baseline for sentence classification tasks. Usage of RNNs for text classification was introduced in [7] and Bi-LSTM was augmented with simple attention in [10]. Classification results of these models on Hindi text are reported in this work.

Sentence embeddings evaluated in this work include multi-lingual LASER embeddings [2] and multi-lingual BERT based embeddings [4]. LASER uses Bi-LSTM encoder to generate embeddings whereas BERT is based on Transformer architecture. LASER takes a neural machine translation approach for learning sentence representations. It builds a sequence to sequence model using Bi-LSTM encoder-decoder architecture. The encoder Bi-LSTM is used to generate sentence representations. BERT, on the other hand, uses bi-directional transformer encoder for learning word and sentence representations. It uses masked language model as the pre-training objective to mitigate the problem of unidirectional training in simple language model next word prediction task.

3 Datasets

– TREC question dataset which involves classifying a question sentence into six types. The dataset has predefined train-test split. It has 5452 training samples and 500 testing samples. 10% of the training data was randomly held out for validation.
– Stanford Sentiment Treebank datasets SST-1 and SST-2. SST-1 contains one sentence movie reviews which are rated in the scale of 1–5 going from positive to negative. The dataset has predefined train-test-dev split. It has 8544 training samples, 2210 testing samples, and 1101 validation samples. SST-2 is a binary version of SST-1 where there are only two labels positive and negative. It has 6920 training samples, 1821 testing samples, and 872 validation samples.

Original English versions of this dataset are translated to Hindi using Google Translate. A language model was trained using Hindi wiki corpus and used to filter out noisy sentences. We assume no out of vocabulary words as fast text model generates word embeddings for unknown words as well. A common vocabulary of 31k words is created and fast-text vectors are used to initialize the embedding matrix.

4 Model Architectures

The data samples comprise of a sequence of words so different sequence processing models are explored in this work. While the most natural sequence processing model is LSTM, other models are equally applicable as the sequence length is short.

– **BOW**: The bag of words model does not consider the sequence of words. The word vectors of input sentence are averaged to get a sentence embedding of size 300. This is followed by a dense layer of size equal to the number of output classes. Softmax output is given to cross-entropy loss function and Adam is used as an optimizer.
– **BOW + Attention**: In this model, instead of simply averaging, a weighted average of word vectors is taken to generate sentence embedding. The size of sentence embedding is 300 and is followed by a dense layer similar to BOW model. The weights for the individual time step is learned by passing the corresponding word vector through a linear layer of size 300×1. Softmax over these computed weights gives the probabilistic attention scores. This attention approach is described in [10].

- **CNN**: The sequence of word embeddings are passed through three 1D convolutions of kernel sizes 2, 3, and 4. Each convolution uses a filter size of 128. The output of each of the 1-D convolution is max pooled over time and concatenated to get the sentence representation. The size of this sentence representation is 384 dimensions. There is a final dense layer of size equal to the number of output classes.
- **LSTM**: The word vectors are passed as input to two-layer stacked LSTM. The output of the final time step is given as an input to a dense layer for classification. LSTM cell size is 128 and the size of final time step output which is treated as sentence representation is 128.
- **Bi-LSTM**: The sequence of word embedding is passed through two stacked bi-directional LSTM. The output is max pooled over time and followed by a dense layer of size equal to the number of output classes. LSTM cell size is 128 and the size of max-pooled output which is treated as sentence representation is 256.
- **CNN + Bi-LSTM**: The sequence of word embeddings are passed through a 1D convolution of kernel size 3 and filter size 256. The output is passed through a bi-directional LSTM. The output of Bi-LSTM is max pooled over time and followed by a final dense layer.
- **Bi-LSTM + Attention**: This is similar to Bi-LSTM model. The difference is that instead of max-pooling over the output of Bi-LSTM an attention mechanism is employed as described above.
- **LASER and BERT**: Single pre-trained model for learning multilingual sentence representations in the form of BERT and LASER was released by Google and Facebook respectively. BERT is a 12 layer transformer based model trained on multilingual data of 104 languages. LASER a 5 layer Bi-LSTM model pre-trained on multilingual data of 93 languages. Both of these models have Hindi as one of the training languages. The sentence embeddings extracted from these models are used without any fine-tuning or modifications. The pre-trained sentence embeddings are extracted from the corresponding models and subjected to a dense layer of 512 units. It is further connected to a dense layer of size equal to the number of output classes over which softmax is computed. BERT generated 768-dimensional embedding whereas the dimension of LASER embeddings were 1024.

5 Results and Discussion

Performance of different models based on CNN and LSTM were evaluated on translated versions of TREC, SST-1, and SST-2 datasets. Different versions of input word vectors were given to the models for comparison. Pre-trained fast text embeddings trained on Hindi corpus were compared against random initialization of word vectors. The random values were sampled from a continuous uniform distribution in a half-open interval [0.0, 1.0). Moreover, in one setting pre-trained fast-text embeddings were fine-tuned whereas in other settings they remained static. Keeping the word vector layer un-trainable allows better handling of words that were not seen during training as all the word vectors follow the same distribution. However, the domain of the corpus on which the word vectors were pre-trained may be different from the target domain. In such cases, fine-tuning the trained word vectors helps model adapt to the domain of the target corpus. So re-training the fast text vectors and keeping them static has its pros and cons. Table 1 shows the results of the comparison. The three versions of word vectors are indicated as random for random initialization, fast text for trainable fast-text initialization, and fast text-static for un-trainable fast text initialization. Out of all the models vanilla CNN performs the best for all the datasets. CNNs have known to perform best for short texts and same is visible here as the datasets under consideration do not have long sentences. There is a small difference in the performance of different LSTM model. However, Bi-LSTM with max-pooling performed better than its attention version and unidirectional LSTM. Bag of words based on attention fared better than the simple bag of

Table 1. Classification accuracies of different models

Model/Dataset		TREC	SST-1	SST-2
BOW	Fast text-static	62.4	32.2	63.1
	Fast text	87.2	40.4	77.3
	Random	84.4	39.3	76.9
BOW-Attn	Fast text-static	76.2	37.4	72.8
	Fast text	88.2	39.3	78.0
	Random	86.0	36.9	75.4
LSTM	Fast text-static	86.6	40.2	75.5
	Fast text	87.8	40.8	78.1
	Random	86.8	40.7	76.8
Bi-LSTM	Fast text-static	87.0	40.8	76.4
	Fast text	89.8	41.9	78.0
	Random	87.6	40.2	72.9
Bi-LSTM-Attn	Fast text-static	85.0	39.0	76.4
	Fast text	88.6	40.1	78.6
	Random	86.0	39.5	76.0
CNN	Fast text-static	91.2	41.2	78.2
	Fast text	**92.8**	**42.9**	**79.4**
	Random	87.8	40.2	77.1
CNN+Bi-LSTM	Fast text-static	89.6	40.4	78.3
	Fast text	90.5	41.0	77.4
	Random	87.6	38.2	72.3
LASER		89.0	41.4	75.9
BERT		77.6	35.6	68.5

words model. Attention was particularly helpful with the usage of static fast text word vectors. Stacked CNN-LSTM models were somewhere between LSTM and CNN based models. We did not see a huge drop in performance due to random initialization of word vectors. But the performance across different epochs was very stable with fast text initialization. Finally, as compared to generic sentence embeddings obtained from BERT and LASER, specific embeddings obtained from custom models performed better. LASER was able to reach close to the best performing model. This shows that LASER was able to capture important discriminative features of a sentence required for the task at hand whereas BERT failed to capture the same.

6   Conclusion

In this work, we compared different deep learning approaches for Hindi sentence classification. The word vectors were initialized using fast text word vectors trained on Hindi corpus and random word vectors. This work also serves the evaluation of fast text word embeddings for Hindi sentence classification task. CNN models perform better than LSTM based models on the datasets considered in this paper. Although we would expect BOW to perform the worst it has numbers comparable to LSTM and CNN. Therefore if we can trade off accuracy for speed BOW is useful. LSTMs do not do better than CNNs may be because the word order is relaxed in Hindi. Sentence representations captured by LASER multilingual model were rich as compared to BERT. However, overall custom trained models on specific datasets performed better than lightweight models directly utilizing sentence encodings. Although the real advantage of multi-lingual embeddings can be better evaluated on tasks involving text from multiple languages.

References

1. Arora, P.: Sentiment analysis for hindi language (2013)
2. Artetxe, M., Schwenk, H.: Massively multilingual sentence embeddings for zero-shot cross-lingual transfer and beyond. arXiv preprint arXiv:1812.10464 (2018)
3. Conneau, A., Lample, G., Rinott, R., Williams, A., Bowman, S.R., Schwenk, H., Stoyanov, V.: Xnli: evaluating cross-lingual sentence representations. arXiv preprint arXiv:1809.05053 (2018)
4. Devlin, J., Chang, M.W., Lee, K., Toutanova, K.: Bert: pre-training of deep bidirectional transformers for language understanding. arXiv preprint arXiv:1810.04805 (2018)
5. Joulin, A., Grave, E., Bojanowski, P., Mikolov, T.: Bag of tricks for efficient text classification. arXiv preprint arXiv:1607.01759 (2016)
6. Kim, Y.: Convolutional neural networks for sentence classification. arXiv preprint arXiv:1408.5882 (2014)
7. Lai, S., Xu, L., Liu, K., Zhao, J.: Recurrent convolutional neural networks for text classification. In: Twenty-ninth AAAI Conference on Artificial Intelligence (2015)

8. Pennington, J., Socher, R., Manning, C.: Glove: global vectors for word representation. In: Proceedings of the 2014 Conference on Empirical Methods in Natural Language Processing (EMNLP), pp. 1532–1543 (2014)
9. Tummalapalli, M., Chinnakotla, M., Mamidi, R.: Towards better sentence classification for morphologically rich languages (2018)
10. Zhou, P., et al.: Attention-based bidirectional long short-term memory networks for relation classification. In: Proceedings of the 54th Annual Meeting of the Association for Computational Linguistics (Volume 2: Short Papers), pp. 207–212 (2016)

A Stacked Ensemble Approach to Bengali Sentiment Analysis

Kamal Sarkar[(✉)]

Computer Science and Engineering Department, Jadavpur University, Kolkata 700032, India
jukamal2001@yahoo.com

Abstract. Sentiment analysis is a crucial step in the social media data analysis. The majority of research works on sentiment analysis focus on sentiment polarity detection which identifies whether an input text is positive, negative or neutral. In this paper, we have implemented a stacked ensemble approach to sentiment polarity detection in Bengali tweets. The basic concept of stacked generalization is to fuse the outputs of the first level base classifiers using a second-level Meta classifier in an ensemble. In our ensemble method, we have used two types of base classifiers- multinomial Naïve Bayes classifiers and SVM that make use of a diverse set of features. Our proposed approach shows an improvement over some existing Bengali sentiment analysis approaches reported in the literature.

Keywords: Bengali tweets · Sentiment polarity detection · Machine learning · Stacked ensemble · Meta classifier

1 Introduction

Now-a-days, a huge amount of social media and user-generated content are available on the Web. The major portion of social media texts such as blog posts, tweets and comments includes opinion related information. The internet users give opinions in the various domains such as politics, crickets, sports, movies, music etc. This vast amount of online social media textual data can be collected and mined for deriving intelligence useful in the many applications in the various domains such as marketing, politics/political science, policy making, sociology and psychology. A social media user expresses the sentiment in form of an opinion, a subjective impression, a thought or judgment prompted by feelings [1]. It also includes emotions. Sentiment analysis for detecting sentiment polarity (positive, negative or neutral) in social media texts has been recognized by the researchers as one of the major research topics though it is hard to find the concrete boundary between two research areas-sentiment analysis and opinion mining.

So, to derive knowledge from the vast amount of social media data, there is a need for an efficient and accurate system which can perform analysis and detect sentiment polarity of the opinions coming from various heterogeneous sources.

The most common approaches to sentiment analysis use various machine learning based techniques [2–9] though the earlier approaches to sentiment analysis used the natural language processing (NLP) and computational linguistics techniques [10–13]

© Springer Nature Switzerland AG 2020
U. S. Tiwary and S. Chaudhury (Eds.): IHCI 2019, LNCS 11886, pp. 102–111, 2020.
https://doi.org/10.1007/978-3-030-44689-5_10

which use in-depth linguistic knowledge. Research works on sentiment polarity detection have already been carried out in the different genres: blogs [14], discussion boards or forums [15], user reviews [16] and expert reviews [17].

In contrast to machine learning based approach to sentiment analysis, the lexicon based approach [18] relies solely on background knowledge base or sentiment lexicon which is either manually constructed lexicon of polarity (positive and negative) terms [19] or a lexicon constructed using some automatic process [20–25]. A kind of sentiment lexicon [26] is manually constructed lexicon created based on the quantitative analysis of the glosses associated with synsets retrieved from WordNet [27]. Though the sentiment lexicon plays a crucial role in the most sentiment analysis tasks, the main problem with the lexicon based approach is that it is difficult to extract and maintain a universal sentiment lexicon.

The main advantage of the machine learning based approach is that it can easily be ported to any domain quickly by changing the training dataset. Hence the most researchers prefer supervised machine based sentiment analysis approach. Supervised method for sentiment polarity detection uses machine learning algorithms trained on a sentiment polarity labeled corpus of social media texts where each text is turned into a feature vector. The most commonly used features are word n-grams, surrounding words, or even punctuations. The most common machine learning algorithms which have been used for sentiment analysis task are Naïve Bayes, SVM, K-nearest neighbor, decision tree, Artificial Neural Networks [18, 28–36].

The most previous research works on sentiment polarity detection involves analysis of sentiments of English texts. But due to the multilingual nature of Indian social media texts, there is also need for developing a system that can do sentiment analysis of Indian language texts. In this line, a shared task on Sentiment Analysis in Indian Languages (SAIL) Tweets was held in conjunction with MIKE 2015 conference at IIIT Hyderabad, India [37]. Bengali language was also included in this shared task as one of the major Indian languages. The Bengali (Bangla) Language is also one of the most spoken languages in the world. In recent years, some researchers have attempted to develop sentiment analysis systems for Bengali language [35, 38–41].

In this paper, we present a stacked ensemble approach for sentiment polarity detection of Bengali tweets. This approach first constructs three base models, where each model makes use of a subset of input features representing the tweets. The tweets are represented by word n-gram, character n-gram and SentiWordNet features. SentiWord-Net or sentiment-WordNet is an external knowledge base containing a collection of polarity words (discussed in the next section).

Features are grouped into three subsets- (1) the subset consisting of word n-gram features and Sentiment-WordNet features, (2) the subset consisting of character n-gram features and Sentiment-WordNet features and (3) the subset consisting of unigram and Sentiment-WordNet features. The first base model is developed using multinomial Naïve Bayes with the word n-gram and Sentiment-WordNet features, the second base model is developed using multinomial Naïve Bayes with character n-gram and Sentiment-WordNet features and the third base model is developed using support vector machine with unigram features and Sentiment-WordNet features. The base level classifiers' predictions are combined using a Meta classifier. This process is popularly known as stacking.

2 Proposed Methodology

The proposed system uses stacked ensemble model for sentiment polarity detection in Bengali tweets. The proposed model has three important steps: (1) data cleaning, (2) features and base classifiers (3) model development and sentiment polarity classification.

2.1 Data Cleaning

At the preprocessing step, the entire data collection is processed to remove irrelevant characters from the data. This is important for tweet data because tweet data is noisy.

2.2 Features and Base Classifiers

Our idea of stacked ensemble is to combine outputs of the base classifiers, each of which makes use of a subset of features representing the tweets. As mentioned earlier, a tweet is represented by word n-gram, character n-gram features and SentiWordNet features. The word n-grams or the character n-grams that do not occur at least 3 times in the training data are removed from the tweets as noise.

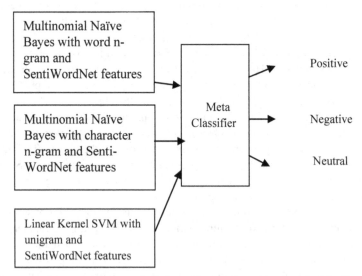

Fig. 1. System architecture of the stacked ensemble based machine learning for Bengali tweet sentiment analysis

We have developed three base classifiers at the first level-(1) multinomial naïve Bayes with word n-gram features and Sentiment-WordNet features, (2) multinomial naïve

Bayes with character n-gram features and Sentiment-WordNet features, and (3) linear kernel support vector machines with unigram (1-gram) features and Sentiment-WordNet features. For the second level, we have used MLP classifier as the Meta classifier. The overall architecture of our proposed model is shown in Fig. 1.

Base Classifiers

Multinomial Naïve Bayes with word n-gram features and Sentiment-WordNet features. As mentioned earlier, the first base model in our proposed stacked ensemble model uses multinomial naïve Bayes [38, 40] and the associated feature set includes word n-gram features and Sentiment-WordNet features. For this model, we have taken unigrams, bigrams and trigrams as the features (we have taken up to trigrams, i.e., n = 1, 2, 3). Word n-grams which do not occur at least 3 times in the training data are removed as noise. Considering word *n-gram* as features, the sentiment class of a tweet T is determined by the posterior probability for a sentiment class given the sequence of word n-grams in the tweet:

$$P(C|T) = P(C) \prod_{i=1}^{m} P(t_i|C) \tag{1}$$

Where: m is the number of word n-grams (n = 1 to 3) in the tweet,
t_i is the i-th word n-gram type,
C is a sentiment class,
P(C) is the prior probability,
T is a tweet represented as a sequence of word n-grams in the tweet, $T = (t_1, t_2, \ldots t_m)$ and
m is the number of word n-grams in the tweet including repetition (a word n-gram may repeat in the tweet).

The details of how the Multinomial Naïve Bayes is applied to sentiment analysis task can be found in [38]. In addition to word n-gram features, we have also incorporated external knowledge base called Sentiment-Wordnet wherefrom some polarity information is retrieved for the tweet words. Though polarity of a word does not always depend on its literal meaning, there are many words which are usually used as positive words (for example, the word "good"). This is also true for negative polarity words. Such information may be useful for sentiment polarity detection in tweets. For our work, Sentiment-WordNet for Indian Languages [42] (retrieved from http://amitavadas.com/sentiwordnet.php) has been used. This is a collection of positive, negative and neutral words along with their broad part-of-speech categories. To incorporate Sentiment-WordNet, each word in a tweet of the corpus is augmented with a special word "#P" if the tweet word is found in the list of positive polarity words, "#N" if the tweet word is found in the negative set and "#NU" if the word is found in the neutral set. For example, the tweet " এটা একটা খুবই ভালো থাবার" (This is a very good food) is augmented as follows: " এটা একটা খুবই#P ভালো#P থাবার" (This is a very #P good #P food). With this new augmentation, the formula for posterior probability is modified as follows:

$$P(C|T) = P(C) \left[\prod_{i=1}^{m} P(t_i|C) \right] P(\#P|C)^{m1} P(\#N|C)^{m2} P(\#NU|C)^{m3} \tag{2}$$

Where: m = number of word n-grams (n = 1 to 3) in the tweet (including repetition)

m1 = number of tweet words found in the positive word-list of Sentiment-WordNet.
m2 = number of tweet words found in the negative word-list of Sentiment-WordNet.
m3 = number of tweet words found in the neutral-word list of Sentiment-WordNet.

From Eq. 2, it is evident that the posterior probability for a tweet is boosted by how many polarity words it contains. For example, if a tweet contains more number of positive polarity words than other two types, the overall polarity of the tweet is boosted in the direction of positivity.

Multinomial Naïve Bayes with Character n-gram Features and Sentiment-WordNet Features. This base classifier is also based on the same principle described in the above sub-section. The only difference is that this model makes use of the different subset of input features, that is, it uses character *n*-grams and Sentiment-WordNet features representing a tweet. The examples of character *n-grams* and *word n-grams* are given below:

Example Input text: *"khub bhalo cinema"* (very good movie).

Word *n*-grams (for n = 1, 2) are: "khub", "bhalo", "cinema", "khub bhalo", "bhalo cinema".

Character *n*-grams for n = 4 are: "khub", "hub", "ub b", "b bh", "bha", "bhal", "halo", "alo", "lo c", "o cin", "cin", "cine", "inem", "nema".

It is very common that the word occurring in the test tweet is absent in the training data. This is known as out-of-vocabulary problem. Character *n*-gram features are useful to deal with the out-of-vocabulary problem. The character *n*-grams with n varying from 2 to 5 are used for developing this base model. The character n-grams that do not occur at least 3 times in the training data are removed as noise.

For this base model, the set of character *n*-gram features and the Sentiment-WordNet features are used and the posterior probability for the tweet is calculated using Eq. 2 with the only difference is that the variables $t_1, t_2, \ldots t_m$ in Eq. 2 refer to the distinct character n-grams in the tweet, that means, the probability value is taken only once in the equation even if the character n-gram repeats several times in the tweet.

Support Vector Machines with Unigram and Sentiment-WordNet Features. It is proven that Support Vector Machines (SVM) [43] with linear kernel is useful in text classification task due to its inherent capability in dealing with high dimensionality of the data. So, for the third base classifier, SVM with linear kernel has been used. This base model also uses a different subset of tweet features, that is, it uses unigram (word 1-gram) features and Sentiment-WordNet features. Since the tweet words, which are found in Sentiment-WordNet, are augmented with one of the possible pseudo words - "#P", "#N" and "#NU", Sentiment-WordNet features are automatically taken into account while computing the unigram feature set.

For developing this base model, we did not take all unigrams as features. A subset of unigrams is taken as features because we observe that increasing the number of unigram features hampers the individual performance of this base model. So, for this purpose, the most frequent 1000 unigrams per class are considered as the features for developing this base classifier. Thus, according to bag-of-unigrams model, each tweet is represented by a feature vector of length 3000 (1000 per class × 3) where each component of the vector corresponds to the frequency of the corresponding unigram in the tweet under consideration and finally each vector is labelled by the label of the corresponding training tweet.

2.3 Model Development and Sentiment Classification

As we have shown in Fig. 1, for model development, three base classifiers are used and the base classifiers' predictions are combined using a Meta classifier. We have used multilayer perceptron (MLP) neural network classifier as the meta-classifier at the second level. From the training data provided to the model, it learns how to classify a tweet into one of three sentiment polarity classes - Positive, Negative and Neutral. Here the MLP classifier has one hidden layer with *softplus* activation function. The number of nodes considered in the hidden layer is 2.

During testing phase, the unlabeled tweet is presented for classification to the trained model. The label of the test tweet, assigned by the model, is considered as the sentiment label of the corresponding tweet.

3 Evaluation and Experimental Results

We have used Bengali datasets released for a shared task on Sentiment Analysis in Indian Languages (SAIL) Tweets, held at IIIT Hyderabad, India [37]. The training set consists of 1000 tweets and the test set consists of 500 tweets.

3.1 Experiments and Results

We have combined the SAIL training and test data to form a dataset consisting of 1500 tweets and 10-fold cross validation is done and the average accuracy over 10 folds is computed for each model presented in this paper. The obtained average accuracy has been reported in this paper.

We have compared our proposed stacked ensemble model with some existing Bengali tweet sentiment analysis systems published in the literature. For meaningful comparisons among the systems, we implemented the existing systems that previously used SAIL 2015 datasets for system development. The brief description of the systems to which our proposed system is compared is given below.

- A deep learning model for Bengali tweet sentiment analysis has been presented in [41]. It has used recurrent neural networks called LSTM for model development. This LSTM based model takes into account the entire sequence of tokens contained in a tweet while detecting sentiment polarity of the tweet. The similar tweet augmentation strategy used in our proposed model is also used in this model.
- The sentiment polarity detection model in [38] uses Multinomial Naïve Bayes with word unigram, bigram and Sentiment-WordNet features. The details can be found in [38].
- The sentiment polarity detection model reported in [38] uses SVM with word unigram and Sentiment-WordNet features. The details of this model can be found in [38].
- The sentiment polarity detection model presented in [40] uses character n-gram features and Sentiment-WordNet features. The details of this model can be found in [40].

Table 1. Performance comparisons of our proposed model and other four existing machine learning based models applied to sentiment polarity detection in Bengali tweets

Models	Accuracy (%)
Our proposed stacked ensemble model	56.26
Deep learning (LSTM) based model [41]	55.27
Multinomial Naïve Bayes based model with character n-gram and SentiWordNet features [40]	55.2
SVM based model [38]	53.73
Multinomial Naïve Bayes based model with word n-gram and SentiWordNet features [38]	53.07

We have compared the results obtained by our proposed stacked ensemble model with four existing sentiment polarity detection models described above. The comparisons of the results have been shown in Table 1. It is evident from Table 1 that our proposed stacked ensemble models.perform better than other existing models it is compared to. Since each of the existing models mentioned in Table 1 uses a single machine learning algorithm that uses either word n-gram and Sentiment-WordNet features or character n-gram and Sentiment-WordNet features for Bengali tweet sentiment classification, the results obtained by our proposed stacked ensemble model show that combining classifiers with stacking improves performance over the individual classifier applied to sentiment polarity detection in Bengali tweets. As we can see from Table 1, our proposed model also performs better than a LSTM based deep learning model presented in [41].

4   Conclusion and Future Work

In this paper, we have described stacked ensemble model for Bengali tweet sentiment classification. Two Multinomial Naïve Bayes models using the different subsets of features and a SVM based model with linear kernel have been combined in a stacked ensemble using MLP classifier. We have experimented to choose the appropriate Meta classifier and our experiments reveal that MLP classifier with *softplus* activation in the hidden units performs best among other possibilities we considered.

The insufficiency of the training data is one of the major problems for developing systems for Bengali tweet sentiment analysis. We also observe that SAIL 2015 data is not error free. Some tweets have been wrongly labeled by the human annotators. However, for meaningful comparisons of the systems, we have left those errors uncorrected. We hope that the system performance can be improved with the increased amount of training data and proper annotation. We also expect that our proposed system can be easily ported to other Indian languages like Hindi, Tamil etc.

Choosing appropriate base classifiers and the Meta classifier can be the other ways for improving the system performance.

Acknowledgments. This research work has received support from the project entitled *"Indian Social Media Sensor: an Indian Social Media Text Mining System for Topic Detection, Topic Sentiment Analysis and Opinion Summarization"* funded by the Department of Science and Technology, Government of India under the SERB scheme.

References

1. Bowker, J.: The Oxford Dictionary of World Religions. Oxford University Press, Oxford (1997)
2. Zhao, J., Liu, K., Wang, G.: Adding redundant features for CRFs-based sentence sentiment classification. In: Proceedings of the Conference on Empirical Methods in Natural Language Processing, pp. 117–126. Association for Computational Linguistics (2008)
3. Joachims, T.: Making large scale SVM learning practical. In: Schölkopf, B., Burges, C.J.C., der Smola, A. (eds.) Advances in Kernel Methods-Support Vector Learning. MITPress, Cambridge (1999)
4. Pang, B., Lee, L., Vaithyanathan, S.: Thumbs up?: sentiment classification using machine learning techniques. In: Proceedings of the ACL-02 Conference on Empirical Methods in Natural Language Processing, vol. 10, pp. 79–86. Association for Computational Linguistics (2002)
5. Dave, K., Lawrence, S., Pennock, D.M.: Mining the peanut gallery: opinion extraction and semantic classification of product reviews. In: Proceedings of the 12th International Conference on World Wide Web, pp. 519–528. ACM (2003)
6. Mullen, T., Collier, N.: Sentiment analysis using support vector machines with diverse information sources. In: EMNLP, vol. 4, pp. 412–418 (2004)
7. Pang, B., Lee, L.: Opinion mining and sentiment analysis. Found. Trends Inf. Retrieval **2**(1–2), 1–135 (2008)
8. Goldberg, A.B., Zhu, X.: Seeing stars when there aren't many stars: graph-based semi-supervised learning for sentiment categorization. In: Proceedings of the First Workshop on Graph Based Methods for Natural Language Processing, pp. 45–52. Association for Computational Linguistics (2006)
9. Miao, Q., Li, Q., Zeng, D.: Fine grained opinion mining by integrating multiple review sources. J. Am. Soc. Inform. Sci. Technol. **61**(11), 2288–2299 (2010)
10. Riloff, E., Wiebe, J.: Learning extraction patterns for subjective expressions. In: Proceedings of the 2003 Conference on Empirical Methods in Natural Language Processing, pp. 105–112. Association for Computational Linguistics (2003)
11. Prabowo, R., Thelwall, M.: Sentiment analysis: a combined approach. J. Inf. **3**(2), 143–157 (2009)
12. Narayanan, R., Liu, B., Choudhary, A.: Sentiment analysis of conditional sentences. In: Proceedings of the 2009 Conference on Empirical Methods in Natural Language Processing, vol. 1, pp. 180–189. Association for Computational Linguistics (2009)
13. Wiegand, M., Balahur, A., Roth, B., Klakow, D., Montoyo, A.: A survey on the role of negation in sentiment analysis. In: Proceedings of the Workshop on Negation and Speculation in Natural Language Processing, pp. 60–68. Association for Computational Linguistics (2010)
14. Ku, L.-W., Liang, Y.T., Chen, H-H.: Opinion extraction, summarization and tracking in news and blog corpora. In: AAAI Spring Symposium: Computational Approaches to Analyzing Weblogs (2006)
15. Kim, J., Chern, G., Feng, D., Shaw, E., Hovy, E.: Mining and assessing discussions on the web through speech act analysis. In: Proceedings of the Workshop on Web Content Mining with Human Language Technologies at the 5th International Semantic Web Conference (2006)

16. Pang, B., Lee, L.: A sentimental education: sentiment analysis using subjectivity summarization based on minimum cuts. In: Proceedings of the 42nd Annual Meeting on Association for Computational Linguistics, p. 271 (2004)
17. Zhu, F., Zhang, X.: Impact of online consumer reviews on sales: the moderating role of product and consumer characteristics. J. Mark. **74**(2), 133–148 (2010)
18. Melville, P., Gryc, W., Lawrence, R.D.: Sentiment analysis of blogs by combining lexical knowledge with text classification. In: Proceedings of the 15th ACM SIGKDD International Conference on Knowledge Discovery and Data Mining, pp. 1275–1284. ACM (2009)
19. Ramakrishnan, G., Jadhav, A., Joshi, A., Chakrabarti, S., Bhattacharyya, P.: Question answering via Bayesian inference on lexical relations. In: Proceedings of the ACL 2003 Workshop on Multilingual Summarization and Question Answering, vol. 12, pp. 1–10. Association for Computational Linguistics (2003)
20. Jiao, J., Zhou, Y.: Sentiment Polarity Analysis based multi-dictionary. Phys. Procedia **22**, 590–596 (2011)
21. Macdonald, C., Ounis, I.: The TREC Blogs06 collection: creating and analysing a blog test collection. Department of Computer Science, University of Glasgow Technical report TR-2006-224, 1, 3-1, (2006)
22. Hatzivassiloglou, V., McKeown, K.R.: Predicting the semantic orientation of adjectives. In: Proceedings of the 35th Annual Meeting of the Association for Computational Linguistics and Eighth Conference of the European Chapter of the Association for Computational Linguistics, pp. 174–181. Association for Computational Linguistics, July 1997
23. Wiebe, J.: Learning subjective adjectives from corpora. In: AAAI/IAAI, pp. 735–740, July 2000
24. Yu, H., Hatzivassiloglou, V.: Towards answering opinion questions: separating facts from opinions and identifying the polarity of opinion sentences. In: Proceedings of the 2003 Conference on Empirical Methods in Natural Language Processing, pp. 129–136. Association for Computational Linguistics, July 2003
25. Riloff, E., Wiebe, J.: Learning extraction patterns for subjective expressions. In: Proceedings of the 2003 Conference on Empirical Methods in Natural Language Processing, pp. 105–112. Association for Computational Linguistics, July 2003
26. Esuli, A., Sebastiani, F.: SENTIWORDNET: a publicly available lexical resource for opinion mining. In: Proceedings of LREC, vol. 6, pp. 417–422, May 2006
27. Fellbaum, C.: WordNet. Blackwell Publishing Ltd., Hoboken (1999)
28. Pang, B., Lee, L.: A sentimental education: sentiment analysis using subjectivity summarization based on minimum cuts. In: ACL (2004)
29. Chen, C.C., Tseng, Y.D.: Quality evaluation of product reviews using an information quality framework. Decis. Support Syst. **50**(4), 755–768 (2011)
30. Kang, H., Yoo, S.J., Han, D.: Senti-lexicon and improved Naïve Bayes algorithms for sentiment analysis of restaurant reviews. Expert Syst. Appl. **39**(5), 6000–6010 (2012)
31. Clarke, D., Lane, P., Hender, P.: Developing robust models for favourability analysis. In: Proceedings of the 2nd Workshop on Computational Approaches to Subjectivity and Sentiment Analysis, pp. 44–52. Association for Computational Linguistics (2011)
32. Reyes, A., Rosso, P.: Making objective decisions from subjective data: detecting irony in customer reviews. Decis. Support Syst. **53**(4), 754–760 (2012)
33. Moraes, R., Valiati, J.F., Neto, W.P.G.: Document-level sentiment classification: an empirical comparison between SVM and ANN. Expert Syst. Appl. **40**(2), 621–633 (2013)
34. Martín-Valdivia, M.T., Martínez-Cámara, E., Perea-Ortega, J.M., Ureña-López, L.A.: Sentiment polarity detection in Spanish reviews combining supervised and unsupervised approaches. Expert Syst. Appl. **40**(10), 3934–3942 (2013)

35. Sarkar, K., Chakraborty, S.: A sentiment analysis system for Indian Language Tweets. In: Prasath, R., Vuppala, A.K., Kathirvalavakumar, T. (eds.) MIKE 2015. LNCS (LNAI), vol. 9468, pp. 694–702. Springer, Cham (2015). https://doi.org/10.1007/978-3-319-26832-3_66

36. Li, Y.M., Li, T.Y.: Deriving market intelligence from microblogs. Decis. Support Syst. **55**(1), 206–217 (2013)

37. Patra, B.G., Das, D., Das, A., Prasath, R.: Shared task on Sentiment Analysis in Indian Languages (SAIL) tweets - an overview. In: Prasath, R., Vuppala, A.K., Kathirvalavakumar, T. (eds.) MIKE 2015. LNCS (LNAI), vol. 9468, pp. 650–655. Springer, Cham (2015). https://doi.org/10.1007/978-3-319-26832-3_61

38. Sarkar, K., Bhowmik, M.: Sentiment polarity detection in bengali tweets using multinomial Naïve Bayes and support vector machines. In: CALCON 2017, Kolkata. IEEE (2017)

39. Sarkar, K.: Sentiment polarity detection in Bengali tweets using deep convolutional neural networks. J. Intell. Syst. **28**(3), 377–386 (2018). https://doi.org/10.1515/jisys-2017-0418. Accessed 7 July 2019

40. Sarkar, K.: Using character N gram features and multinomial Naïve Bayes for sentiment polarity detection in Bengali tweets. In: Proceedings of Fifth International Conference on Emerging Applications of Information Technology (EAIT), Kolkata. IEEE (2018)

41. Sarkar, K.: Sentiment polarity detection in Bengali tweets using LSTM recurrent neural networks. In: Proceedings of Second International Conference on Advanced Computational and Communication Paradigms (ICACCP), Sikkim, India, 25–28 February 2019. IEEE (2019)

42. Das, A., Bandyopadhyay, S.: SentiWordNet for Indian languages. In: Proceedings of 8th Workshop on Asian Language Resources (COLING 2010), Beijing, China, pp. 56–63 (2010)

43. Vapnik, V.: Estimation of Dependences Based on Empirical Data, vol. 40. Springer-Verlag, New York (1982). https://doi.org/10.1007/0-387-34239-7

Computing with Words Through Interval Type-2 Fuzzy Sets for Decision Making Environment

Rohit Mishra$^{(\boxtimes)}$, Santosh Kumar Barnwal , Shrikant Malviya ,
Varsha Singh , Punit Singh , Sumit Singh , and Uma Shanker Tiwary

Indian Institute of Information Technology Allahabad, Allahabad, India
rohit129iiita@gmail.com, iis2009002@gmail.com, s.kant.malviya@gmail.com ,
thepunitsingh@gmail.com

Abstract. Interval Type-2 fuzzy sets (IT2FSs) are used for modeling
uncertainty and imprecision in a better way. In a conversation, the information given by humans are mostly words. IT2FSs can be used to provide
a suitable mathematical representation of a word. The IT2FSs can be
further processed using Computing with the words (CWW) engine to
return the IT2FS output representation that can be decoded to give the
output word. In this paper, an attempt has been made to develop a system that will help in decision making by considering person's subjective
importance for various factors for selection. For demonstration we have
taken an example of restaurant recommender system that suggests the
suitability of a restaurant depending on person's subjective importance
given to selection criteria (i.e.,*cost, time and food quality*). Firstly, a
codebook is constructed to capture the vocabulary words. IT2FSs membership functions are used to represent these vocabulary words. The linguistic ratings corresponding to selection criteria are taken from experts
for restaurants. The linguistic weights are person's subjective importance
given to the selection criteria. Finally, the CWW engine uses linguistic
weights and linguistic ratings to obtain the suitability of the restaurant.
The output is the recommended word which is also represented using
IT2FS. The output word is more effective for human understanding in
conversation where the precise information is not very useful and sometimes deceptive.

Keywords: Computing with words · Interval Type 2 Fuzzy sets ·
Perceptual computing · Natural language understanding · Dialogue
systems

1 Introduction

Computing with words (CWW) as proposed by L.A. Zadeh, is "a methodology in which words are used in place of numbers for computing and reasoning"
[1,2]. CWW is helpful when dealing with the imprecise information carried by

© Springer Nature Switzerland AG 2020
U. S. Tiwary and S. Chaudhury (Eds.): IHCI 2019, LNCS 11886, pp. 112–123, 2020.
https://doi.org/10.1007/978-3-030-44689-5_11

words. Often the information carried by words have two types of uncertainties, the intra-level and the inter-level uncertainties [3,4]. The intra-level uncertainty about a word is the uncertainty that a subject has about the word. The inter-level uncertainty about a word that is varied among the group of subjects. To tackle this problem, Type-1 fuzzy sets (T1FSs) are not considered suitable to capture the uncertainties of words. Interval Type-2 fuzzy sets (IT2FSs) can be used to capture both types of uncertainties. [5,6]. In this paper, the words associated with selection criteria(i.e., cost, time and food quality) for restaurants have the uncertainties associated with them which are represented using IT2FS. In future, this work can be extended from textual domain to speech domain where uncertainties associated with prosodic features can be easily captured using IT2FS [7,8].

The paper is organised as follows: In Sect. 2, introduction to IT2FS foundation knowledge is introduced. The basic terminologies and notations are explained. In Sect. 3, the details of Methodology to select IT2FS rules are given and it is demonstrated that CWW using IT2FSs can be successfully applied in restaurant selection problem. Results and Discussion is included in Sect. 4.

2 Interval Type-2 Fuzzy Set

In IT2FS $\tilde{A}$ is characterised by membership function (MF) $\mu_{\tilde{A}}(x, u)$, where $x \in X$ and $u \in J_x \subseteq [0, 1]$, that is

$$\tilde{A} = \{((x, u), \mu_{\tilde{A}}(x, u) = 1) | \forall x \in X, \forall u \in J_x \subseteq [0, 1]\} \tag{1}$$

As per description in [12,14], it can be formulated as:

$$\tilde{A} = \int_{x \in X} \int_{u \in J_x \subseteq [0,1]} 1/(x, u) = \int_{x \in X} \left[\int_{u \in J_x \subseteq [0,1]} 1/u \right] \Big/ x \tag{2}$$

where x is *primary variable* and u is *secondary variable*. J_x is called *primary membership* of x. The term inside the large bracket in Eq. (2) is called the *secondary membership function*.

Some terminologies used in IT2FS are as follows:

1. *Footprint of uncertainty* (FOU): It is the union of all the primary membership of $\tilde{A}$

$$FOU(\tilde{A}) = \bigcup_{\forall x \in X} J_x = \{(x, u) : u \in J_x \subseteq [0, 1]\} \tag{3}$$

2. *Upper membership function* (UMF): Denotes the upper bound of FOU($\tilde{A}$) and is represented as $\bar{\mu}_{\tilde{A}}(x)$

$$UMF(\tilde{A}) = \bar{\mu}_{\tilde{A}}(x) = \overline{FOU(\tilde{A})} \qquad \forall x \in X \tag{4}$$

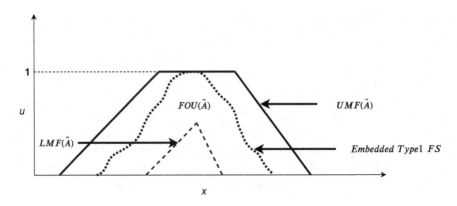

Fig. 1. Interval Type-2 Fuzzy set (IT2FS) [12]

3. *Lower membership function* (LMF): Denotes the lower bound of FOU(Ã) and is represented as $\underline{\mu}_{\tilde{A}}(x)$

$$LMF(\tilde{A}) = \underline{\mu}_{\tilde{A}}(x) = \underline{FOU(\tilde{A})} \qquad \forall x \in X \tag{5}$$

4. *Embedded T1FS A_e*: The set A_e is embedded in FOU(Ã). Equation (6) shows the set of embedded T1FS in continuous space. An example of an *Embedded T1FS* is given in Fig. 1.

$$A_e = \int_{x \in X} u/x \qquad u \in J_x \tag{6}$$

The UMF and LMF are also examples of embedded T1FSs.

5. *Support of $LMF(\tilde{A})$* is the crisp set of all points $x \in X$ such that $LMF(\tilde{A}) > 0$. Similarly *Support of $UMF(\tilde{A})$* is the crisp set of all points $x \in X$ such that $UMF(\tilde{A}) > 0$. The *support* of Ã is the same as the support of $UMF(\tilde{A})$.

For continuous interval, the primary membership J_x is denoted by:

$$J_x = [\underline{\mu}_{\tilde{A}}(x), \bar{\mu}_{\tilde{A}}(x)] \tag{7}$$

Using Eq. (7), the *FOU(Ã)* in Eq. (3) is written as

$$FOU(\tilde{A}) = \bigcup_{\forall x \in X} [\underline{\mu}_{\tilde{A}}(x), \bar{\mu}_{\tilde{A}}(x)] \tag{8}$$

As a consequence, IT2FS can also be represented as:

$$\tilde{A} = 1/FOU(\tilde{A}) \tag{9}$$

It shows that the secondary membership equals 1 for all elements in *FOU(Ã)*. Therefore IT2FS can only be explained by its FOU.

3 Methodology and Experiment

The CWW task is performed on the restaurant recommender system. Here four types of restaurants: *(R1) Posh Restaurants, (R2) Mall Restaurants,(R3) Normal Restaurants* and *(R4) Road side small canteens ('Dhabas')* are considered with three criteria for selection, namely *Cost, Time* and *Food Quality*. These categories are considered based on the data from the expert. Various linguistic ratings for the above selection criteria are mentioned in Fig. 2. In the experiment, three types of the subjects, i.e. *Rich, Middle Income* and *Poor person* are considered with linguistic weights shown in Fig. 3.

Restaurant Category	Selection Criteria		
	Cost (C)	Time (T)	Food Quality (F)
(R1) Posh Restaurants	VH	L	VG
(R2) Mall Restaurants	H	M	G
(R3) Normal Restaurants	LH	LH	F
(R4) Road side small canteens('Dhabas')	VL	L	B

Fig. 2. Linguistic ratings representing the selection criteria for various restaurants

The goal of the system is to suggest the suitability of restaurant for different persons depending on their linguistic weights assigned to different selection criteria. As the input linguistic weights given by the person is expressed in words hence the expected output suggested to the person should be a word. Hence the system should suggest the suitable word for each restaurant. The answer in the word form is considered more suitable to be understood by humans.

In order to develop such a system the methodology requires the following steps to be performed:

Step 1: Data Collection: It includes the collection of *interval endpoints* of the linguistic terms corresponding to a linguistic variable used in the system.
Linguistic terms chosen for corresponding linguistic variables are as follows:

1. **Cost:** *'Very Less (VL)', 'Less (L)', 'Fair (F)', 'Little High (LH)', 'High (H)', 'Very High (VH)'*.
2. **Time:** *'Very Low (VL)', 'Low (L)', 'Little Low (LL)', 'Medium (M)', 'Little High (LH)', 'High (H)', 'Very High (VH)'*.

	Preferences for the 'Selection Criteria'		
Person	**C (Cost)**	**T (Time)**	**F (Food Quality)**
Rich Person	U	VI	VI
Middle Income Person	MLI	MLU	MLI
Poor Person	VI	U	U

Fig. 3. Linguistic weights for specific types of persons

3. **Food Quality:** *'Bad (B)', 'Somewhat Bad (SB)', 'Fair (F)', 'Good (G)', 'Very Good (VG)'.*
4. **Recommended words:** *'Very Low', 'Low', 'Little Low', 'Medium', 'Little High', 'High', 'Very High', 'Extreme'.*

The range for chosen intervals is between [0–10]. The data was collected from 30 randomly selected people with age group between 20 to 30. For example, one of the subject's interval entry for a linguistic term *'Very Less (VL)'* was [0.5,1.9].

Step 2: Interval Pruning: It involved four steps as defined in [5]:

(a) **Bad Data removal:** In this step, the inconsistent data intervals were removed. The following consistency criteria was used (10).

$$
\left.
\begin{array}{l}
0 \le a^{(i)} \le 10 \\
0 \le b^{(i)} \le 10 \\
b^{(i)} \ge a^{(i)}
\end{array}
\right\}, \qquad i = 1, ..., n
\tag{10}
$$

Here, $a^{(i)}$ represents the starting interval point and $b^{(i)}$ represents the ending interval point of ith person.

(b) **Outlier Processing:** Box and Whisker test [15] uses the first, third and inter-quartile range for outlier processing. It was also applied to interval length $L^{(i)} = b^{(i)} - a^{(i)}$. Data interval was accepted when it was satisfying the following conditions or rejected otherwise.

$$
\left.
\begin{array}{l}
a^{(i)} \in [Q_a(0.25) - 1.5IQR_a, Q_a(0.75) + 1.5IQR_a] \\
b^{(i)} \in [Q_b(0.25) - 1.5IQR_b, Q_b(0.75) + 1.5IQR_b] \\
L^{(i)} \in [Q_L(0.25) - 1.5IQR_L, Q_L(0.75) + 1.5IQR_L]
\end{array}
\right\}, \qquad i = 1, ..., n'
$$

$$\tag{11}$$

where $Q_a(0.25)$, $Q_b(0.25)$ and $Q_L(0.25)$ are denoting first quartile of start-point, end-point and length of the interval. Similarly, $Q_{[*]}(0.75)$ is for third quartile and $IQR_{[*]}$ is inter-quartile range. And n' represents the number of remaining intervals after pruning from the last step.

(c) **Tolerance Limit Processing:** For normal distribution the tolerance measurement [15] is given by $m_l \pm k s_l$. The data intervals satisfying the following criteria as in Eq. 12, were accepted. In our case, the value of k was choosen as 2.208 for 95% confidence of the subject intervals.

$$\left.\begin{array}{l} a^{(i)} \in [m_l - ks_l, m_l + ks_l] \\ b^{(i)} \in [m_r - ks_r, m_r + ks_r] \\ L^{(i)} \in [m_L - ks_L, m_L + ks_L] \end{array}\right\}, \qquad i = 1, ..., m' \qquad (12)$$

where m_L and s_L are sample mean and standard deviation of the interval length; m_l and s_l are sample mean and standard deviation of left endpoints of the interval. Similarly, m_r and s_r are sample mean and standard deviation of right endpoints of the interval.

(d) **Reasonable-Interval Processing:** Reasonable intervals are the overlapping intervals. It means that interval range $[a^{(i)}, b^{(i)}]$ should overlap with other intervals. If the data interval pass the following test as defined in [5] then the interval was accepted.

$$\left.\begin{array}{l} a^{(i)} < \xi^* \\ b^{(i)} > \xi^* \end{array}\right\}, \qquad \forall i = 1, ..., m'' \qquad (13)$$

where ξ^* is estimated by:

$$\xi^* = \frac{(m_r \sigma_l^2 - m_l \sigma_r^2) \pm \sigma_l \sigma_r [(m_l - m_r)^2 + 2(\sigma_l^2 - \sigma_r^2) ln(\sigma_l / \sigma_r)]^{1/2}}{\sigma_l^2 - \sigma_r^2} \qquad (14)$$

such that

$$m_l \leq \xi^* \leq m_r \qquad (15)$$

Step 3: Compute Statistics for the Intervals: For each interval $[a^{(i)}, b^{(i)}]$, a uniform probability distribution was chosen. The mean and standard deviation were computed as follows:

$$m^{(i)} = \frac{a^{(i)} + b^{(i)}}{2} \qquad (16)$$

$$\sigma^{(i)} = \frac{b^{(i)} - a^{(i)}}{\sqrt{12}} \qquad (17)$$

These data statistics were mapped to T1FS parameters in Step 4.

Step 4: Creation of Interval Type-2 Fuzzy Sets (IT2FSs): The step consisted of the following substeps: (i) 'Choosing the T1FS model', (ii) 'Computing the uncertainty measure for T1FS model', (iii) 'Establishing Nature of FOU' [5]. The symmetrical triangular T1MF was chosen for interior. For left-shoulder and right-shoulder triangular T1MF of triangle were used in 'Establishing Nature of FOU'. The intervals obtained from Step 3 were classified to one of these three MFs. It gave the set of embedded T1FSs, and by union of these the Interval Type-2 Fuzzy set (IT2FS) was obtained.

The output of Step 4 was an IT2FS codebook containing all the linguistic terms used in our restaurant recommender system. The following FOUs were needed in the restaurant application:

(a) *FOUs for 'Cost'* (represented in Fig. 4)
(b) *FOUs for 'Time'* (represented in Fig. 5)
(c) *FOUs for 'Food Quality'* (represented in Fig. 6)

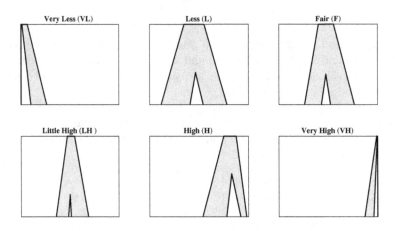

Fig. 4. FOUs for 'Cost'

Step 5: Define the Selection Criteria: It involved defining the linguistic ratings of selection criteria for different categories of restaurants as shown in Fig. 2. Similarly the linguistic weights of persons for selection criteria are shown in Fig. 3.

Step 6: Computing Linguistic Weighted Average (LWA): In this work, CWW Engine uses the *linguistic weighted average* on IT2FSs input to generate IT2FS output. output [9,10,13].

$$\tilde{Y}_{LWA(R_i)} = \frac{\sum_{j=1}^{M} \tilde{X}_{ij}\tilde{W}_j}{\sum_{j=1}^{M} \tilde{W}_j}, \quad i = 1, ..., N \tag{18}$$

In Eq. 18, $\tilde{X}_{ij}$ are IT2FSs related to the linguistic ratings of selection criteria for the restaurant R_i whereas $\tilde{W}_j$ are the linguistic weights corresponding to selection criteria given by a person. N represents the total number of restaurant types and M represents total number of selection criteria for the linguistic weights. Therefore, $\tilde{Y}_{LWA(R_i)}$ represents linguistic weighted average of restaurant R_i.

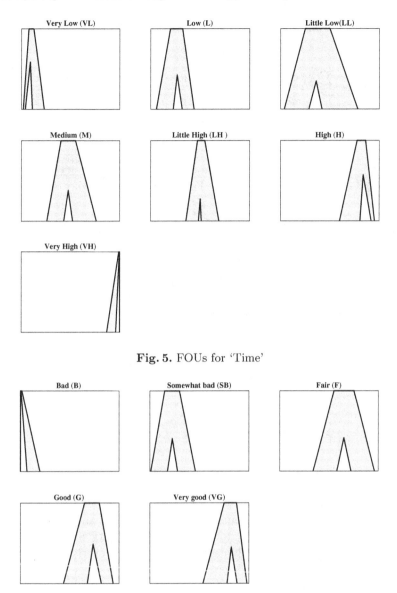

Fig. 5. FOUs for 'Time'

Fig. 6. FOUs for 'Food Quality'

The LWA output $\tilde{Y}_{LWA(R_i)}$ was IT2FS like $\tilde{X}_{ij}$ and $\tilde{W}_j$. As a result of this step, there were N number of $\tilde{Y}_{LWA(R_i)}$, one for each restaurant.

Step 7: Similarity comparison with LWA Output: In the codebook, there was a list of output words with corresponding IT2FSs. The output word suggested was based on *Jaccard similarity measure* [11] between IT2FS of output words in codebook and $\tilde{Y}_{LWA(R_i)}$. The output word was chosen with the highest similarity with $\tilde{Y}_{LWA(R_i)}$.

$$sim_j(\tilde{A}, \tilde{B}) = \frac{\sum_{i=1}^{N} min(\bar{\mu}_{\tilde{A}}(x_i), \bar{\mu}_{\tilde{B}}(x_i)) + \sum_{i=1}^{N} min(\underline{\mu}_{\tilde{A}}(x_i), \underline{\mu}_{\tilde{B}}(x_i))}{\sum_{i=1}^{N} max(\bar{\mu}_{\tilde{A}}(x_i), \bar{\mu}_{\tilde{B}}(x_i)) + \sum_{i=1}^{N} max(\underline{\mu}_{\tilde{A}}(x_i), \underline{\mu}_{\tilde{B}}(x_i))} \quad (19)$$

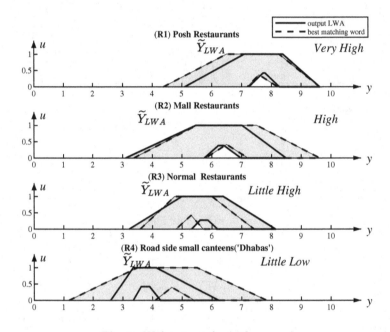

Fig. 7. LWA output for 'rich person'

Where, $\tilde{A}$ is IT2FS for the output word in the codebook and $\tilde{B}$ represents the IT2FS for the LWA output for the restaurant. The similarity $sim_j(\tilde{A}, \tilde{B}) \in [0,1]$, where 1 represents complete overlap and 0 represents no overlap in $\tilde{A}$ and $\tilde{B}$.

4   Results and Discussion

The output of the experiment is shown in Figs. 7, 8 and 9. It shows the LWA output of rich, middle income and poor person corresponding to '(R1) Posh restaurants', '(R2) Mall restaurants', '(R3) Normal restaurants' and '(R4) Road side small canteens (Dhabas)'. The most similar output word in the codebook is mentioned in these figures. The following things can be observed:

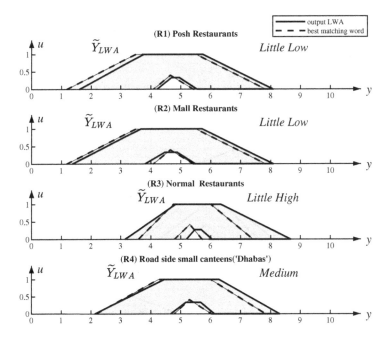

Fig. 8. LWA output for 'middle-class person'

(a). The system suggests (Fig. 7) the following order of preference for the rich person:
'(R1) Posh restaurants' > '(R2) Mall restaurants' > '(R3) Normal restaurants' > '(R4) Road side small canteens(Dhabas)'.
The output word associated with the selection of *'(R1) Posh restaurants'* is *'Very High'*. The output word associated with the selection of *'(R4) Road side small canteens (Dhabas)'* is *'Little Low'* and for other restaurant types the linguistic values range between *'Very High'* and *'Little Low'*.

(b). The system suggests (Fig. 8) the following order of preference for the middle income person:
'(R3) Normal restaurants' > '(R4) Road side small canteens (Dhabas)' > '(R2) Mall restaurants' ≈ '(R1) Posh restaurants'.
The output word associated with the selection of *'(R3) Normal restaurants'* is *'Little High'*. The output word associated with the selection of *'(R1) Posh restaurants'* & *'(R2) Mall restaurants'* is *'Little Low'* and for other restaurant types the linguistic values lies between *'Little High'* and *'Little Low'*.

(c). The system suggests (Fig. 9) the following preference order for the poor person:
'(R4) Road side small canteens (Dhabas)' > '(R3) Normal restaurants' > '(R2) Mall restaurants' > '(R1) Posh restaurants'

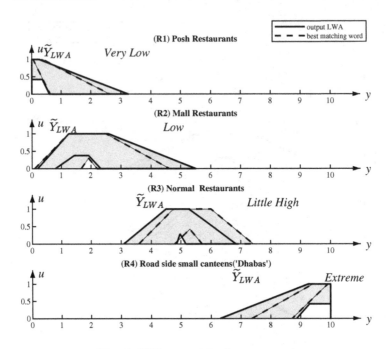

Fig. 9. LWA output for 'poor person'

The output word associated with the selection of various types of restaurant are indicated in the figure itself.

The paper presented a framework for decision making using the CWW (Computing with words) engine. The specific case of selection of restaurant based on different selection criteria of various types of people was handled using IT2FS in implementing a restaurant recommender system. Different types of person have variation in linguistic weights associated with the cost, time and food quality. Similarly, different types of restaurant have variation in linguistic ratings associated with restaurant selection criteria.

The linguistic weights and linguistic ratings carry both inter and intra-levels of uncertainties which are well captured by IT2FSs. The similarity between system output $\tilde{Y}_{LWA}$ of LWA and output words in the codebook was used to suggest the output word for the restaurant to the person. The same framework can also be used to take most appropriate decisions in a different application scenario, such as, selection of appropriate institution for admission, etc.

References

1. Zadeh, L.A.: Fuzzy logic = computing with words. In: Zadeh, L.A., Kacprzyk, J., et al. (eds.) Computing with Words in Information/Intelligent Systems 1, vol. 33, pp. 3–23. Physica, Heidelberg (1999). https://doi.org/10.1007/978-3-7908-1873-4_1

2. Zadeh, L.A.: From computing with numbers to computing with words: from manipulation of measurements to manipulation of perceptions. In: MacCrimmon, M., Tillers, P., et al. (eds.) The Dynamics of Judicial Proof, vol. 94, pp. 81–117. Physica, Heidelberg (2002). https://doi.org/10.1007/978-3-7908-1792-8_5

3. Mendel, J.M., Dongrui, W.: Perceptual reasoning for perceptual computing. IEEE Trans. Fuzzy Syst. **16**(6), 1550–1564 (2008)

4. Wu, H., Mendel, J.M.: Uncertainty bounds and their use in the design of interval type-2 fuzzy logic systems. IEEE Trans. Fuzzy Syst. **10**(5), 622–639 (2002)

5. Liu, F., Mendel, J.M.: Encoding words into interval type-2 fuzzy sets using an interval approach. IEEE Trans. Fuzzy Syst. **16**(6), 1503–1521 (2008)

6. Runkler, T., Coupland, S., John, R.: Interval type-2 fuzzy decision making. Int. J. Approx. Reason. **80**, 217–224 (2017)

7. Mishra, R., Barnwal, S.K., Malviya, S., Mishra, P., Tiwary, U.S.: Prosodic feature selection of personality traits for job interview performance. In: Abraham, A., Cherukuri, A.K., Melin, P., Gandhi, N. (eds.) ISDA 2018 2018. AISC, vol. 940, pp. 673–682. Springer, Cham (2020). https://doi.org/10.1007/978-3-030-16657-1_63

8. Malviya, S., Mishra, R., Tiwary, U.S.: Structural analysis of Hindi phonetics and a method for extraction of phonetically rich sentences from a very large Hindi text corpus. In: 2016 Conference of The Oriental Chapter of International Committee for Coordination and Standardization of Speech Databases and Assessment Techniques (O-COCOSDA). IEEE (2016)

9. Wu, D., Mendel, J.M.: The linguistic weighted average. In: Proceedings of FUZZ-IEEE, pp. 566–573. IEEE, Vancouver (2006)

10. Wu, D., Mendel, J.M.: Aggregation using the linguistic weighted average and interval type-2 fuzzy sets. IEEE Trans. Fuzzy Syst. **15**(4), 1145–1161 (2007)

11. Wu, D., Mendel, J.M.: A comparative study of ranking methods, similarity measures and uncertainty measures for interval type-2 fuzzy sets. Inf. Sci. **179**(8), 1169–1192 (2009)

12. Mendel, J.M., John, R.I., Liu, F.: Interval type-2 fuzzy logic systems made simple. IEEE Trans. Fuzzy Syst. **14**(6), 808–821 (2006)

13. Wu, D., Mendel, J.M.: Corrections to 'aggregation using the linguistic weighted average and interval type-2 fuzzy sets'. IEEE Trans. Fuzzy Syst. **16**(6), 1664–1666 (2008)

14. Mendel, J.M.: Uncertain Rule-Based Fuzzy Logic Systems: Introduction and New Directions. Prentice-Hall, Upper-Saddle River (2001)

15. Walpole, R.W., Myers, R.H., Myers, A.L., Ye, K.: Probability & Statistics for Engineers and Scientists, 8th edn. Prentice-Hall, Upper Saddleback River (2007)

RNN Based Language Generation Models for a Hindi Dialogue System

Sumit Singh[ID], Shrikant Malviya[✉][ID], Rohit Mishra[ID],
Santosh Kumar Barnwal[ID], and Uma Shanker Tiwary[ID]

Indian Institute of Information Technology Allahabad, Allahabad, India
sumitrsch@gmail.com, s.kant.malviya@gmail.com, rohit129iiita@gmail.com,
iis2009002@gmail.com, ustiwary@gmail.com

Abstract. Natural Language Generation (NLG) is a crucial component of a Spoken Dialogue System. Its task is to generate utterances with intended attributes like fluency, variation, readability, scalability and adequacy. As the handcrafted models are rigid and tedious to build, people have proposed many statistical and deep-learning based models to bring about more suitable options for generating utterance on a given Dialogue-Act (DA). This paper presents some Recurrent Neural Network Language Generation (RNNLG) framework based models along with their analysis of how they extract intended meaning in terms of content planning (modelling semantic input) and surface realization (final sentence generation) on a proposed unaligned Hindi dataset. The models have shown consistent performance on our natively developed dataset where the Modified-Semantically-Controlled LSTM (MSC-LSTM) performs better than all in terms of total slot-error (T-Error).

Keywords: Dialogue systems · Natural Language Generation · Recurrent neural network · Delexicalisation · Re-ranking

1 Introduction

In the last decade, substantial research has been done in natural language processing (NLP) specifically in the area of personal assistants such as Alexa, Cortana or Siri. However, the capacity of these conversational agents is still relatively restricted and limited in various ways. One of the most important aspects is the ability to create expressions with a high perceived naturalness and human-like coherence for the content in various domains.

Initially, most NLG systems were based on *rule-based* approaches [1–3] or hybrid of handcrafted and statistical methods [4,5]. As an example, the first statistical NLG model *HALOGEN* was implemented by Langkild et al. which performs reranking on handcrafted candidates using an N-Gram language model (LM) [4]. In 2000, a class-based N-Gram language model (LM) generator, a type of *word-based generator*, was proposed to generate sentences stochastically for a

© Springer Nature Switzerland AG 2020
U. S. Tiwary and S. Chaudhury (Eds.): IHCI 2019, LNCS 11886, pp. 124–137, 2020.
https://doi.org/10.1007/978-3-030-44689-5_12

task-oriented dialogue system [6]. However, inherently it has a very high computation cost during the overgeneration and it is indefinite about covering all the semantics in the outputs. Hence later, the word-based generators were replaced by *phrase-based generators* which had not only reduced the computation cost but also generated linguistically varied utterances [7,8]. However, the phrase-based generators are restricted to semantically-aligned corpora which are tedious and expensive to build.

More recently, researchers have used methods that do not require aligned data and perform *end-to-end* training for getting sentence planning and surface realization done in one go [9]. For achieving the naturalness, variation and scalability on unaligned corpora, they incorporated the deep-learning models. The successful approaches use the RNN-based models to train the *encoder-decoder* on a corpus of paired DAs and corresponding utterances [10,11]. Wen et al. proposed various RNNLG models i.e. Attention-Based Encoder-Decoder (ENC-DEC), Heuristically-gated LSTM (H-LSTM) and Semantically-conditioned LSTM (SC-LSTM) which are also shown to be effective for the NLG module in task-oriented dialogue systems [12,13]. Although, the deep-learning methods are supposed to learn a high-level of semantics [14], but they require a large amount of data for even a small task-oriented system.

Furthermore, in the rule-based and statistical models e.g. N-Gram and KNN Models, the NLG module in an SDS considers only the provided DA as input and cannot adapt to the user's way of speaking. People have tried just not only to avoid the repetition, but also to add variations into the generated responses, typically, either by alternating over a *pool of preset responses* [15], selecting randomly over *k-best generated samples* or using *overgeneration* [12]. The concept of *entrainment* has also been introduced recently into NLG in SDS to enhance the perceived naturalness of the response, but they are mostly rule-based [16]. However, we have observed that none of the approaches have been investigated on a *Hindi-corpora*.

In this paper, we have explored three RNNLG framework[1] based models: (a) H-LSTM, (b) SC-LSTM, (c) MSC-LSTM (Modified SC-LSTM) and (c) ENC-DEC and compared them with the benchmark models i.e. Hand-Crafted (HDC), K-Nearest Neighbour (KNN) model and N-Gram model. All the models are experimented on our own Hindi dialogue dataset, collected on the restaurant domain. The modified RNNLG-models with the proposed dataset are released at the following URL:

https://github.com/skmalviya/RNNLG-Hindi

The paper is organized in five sections including the current section. In Sect. 2, RNNLG framework based models are described. The experimental studies and result & analysis are presented in Sects. 3 and 4 respectively followed by Sect. 5 which mentions conclusion & possible future-extensions.

---

[1] https://github.com/shawnwun/RNNLG.

2   RNNLG Models

This section describes various RNNLG-framework based NLG models. The use of RNN in language generation got inspired after it was successfully applied in sequential data prediction through the RNN based *language modeling* (RNNLM) [17,18]. It was proved that RNNLM learns the output distribution on the previous word as input and generates syntactical correct sequences. However, the sequential output does not ensure the *semantic coherency*. Hence, to generate appropriate sequences, RNN based models are required to be conditioned on semantic-details as well.

Basically in RNNLG, a generator takes DA as an input, comprised of *DA-type* i.e. inform, request, affirm etc., and a set of *slot-value* pairs and generates an appropriate utterance in return. As shown in Figs. 1 and 3, the DA '*inform(name=*राजवाड़ा*, area=*कर्नल गंज*)*' is given as input and the generated output is 'राजवाड़ा कर्नल गंज में है।'. The framework comprises of two parts: first a DA-cell which deals with *content-planning* (semantics) which updates the DA-vector[2] at each time-step stochastically or heuristically and second an s2s (Sequence2Sequence) cell that deals with *surface-realization* (utterance-generation) and update the hidden-vector according to delexicalised utterance as well as current DA-vector as in Eq. 1.

Before training an s2s-cell, the data is first *delexicalised* by replacing the values of slots with specified slot-tokens[3], to make efficient use of training data. The network output is a sequence of tokens that can be lexicalised for the appropriate surface realization.

Typically, a model in RNNLG-framework takes *word2vec* embedding $\mathbf{w}_t$ of a token w_t as an input at each time step t in conditioned with the previous hidden state $\mathbf{h}_{t-1}$ and update it as $\mathbf{h}_t$ for the next iteration to predict the next token w_{t+1} cf., Equations 2 and 3. Furthermore, DA's encoding $\mathbf{d}_t$ is also encapsulated in the RNN in order to embed DA's effect in the generation. Hence, the recurrent function f that updates the hidden state, is represented as:

$$\mathbf{h}_t = f(\mathbf{w}_t, \mathbf{h}_{t-1}, \mathbf{d}_t) \tag{1}$$

This updated hidden state is then transformed to an output probability distribution, from which the next token in the sequence is sampled by function f with softmax-encoding as in Eq. 2. After training the model at each time step t the output is generated through beam-search decoding as in Eq. 3.

$$P(w_{t+1}|w_t, w_{t-1}, ..., w_0, \mathbf{d}_t) = softmax(\mathbf{W}_{ho}\mathbf{h}_t) \tag{2}$$

$$w_{t+1} \sim P(w_{t+1}|w_t, w_{t-1}, ..., w_0, \mathbf{d}_t). \tag{3}$$

---

[2] DA-vector is a 1-hot encoded vector of action-type and slot-value-type where values are corresponding to occurrences of a given slot e.g. sv.name._1, sv.name._2.

[3] Here, token is used to represent both word and slot-token e.g. SLOT_NAME, SLOT_AREA etc. in a delexicalised sentence.

All the RNNLG framework based models are trained corresponding to loss function through Eq. 4:

$$\mathcal{L}(\theta) = \sum_t \mathbf{p}_t^T log(\mathbf{y}_t) + ||\mathbf{d}_T|| + \sum_{t=0}^{T-1} \eta \xi^{||\mathbf{d}_{t+1} - \mathbf{d}_t||} \tag{4}$$

Where, $\mathcal{L}(\theta)$ corresponds to cross-entropy error, θ is training weight-matrix, $\mathbf{d}_T$ is DA-vector of previous time-step. η and ξ are regularization constants (set to 10^{-4} and 100 respectively).

Further, we have explored various ways to implement the recurrent function f such as H-LSTM, SC-LSTM, modified semantically-controlled (MSC-LSTM) and ENC-DEC [13]. All the models follow a common two-cell architecture as mentioned earlier, first a *DA-cell* to model the semantic-input $\mathbf{d}_t$ and second an *LSTM-cell* for updating the hidden-vector $\mathbf{h}_t$.

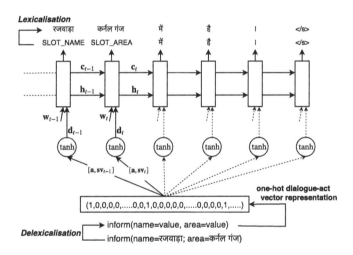

Fig. 1. The architecture of H-LSTM model.

2.1   Heuristically-Gated Model (H-LSTM)

The heuristic approach for DA-cell was proposed to incorporate the current DA-information $\mathbf{d}_t$ in the form of 1-hot encoding vector of DA-type and slot-value pairs in order to generate meaningful utterances [12]. The architecture of H-LSTM model is shown in Fig. 1. In order to avoid the undesirable repetitions in the generation, $\mathbf{d}_t$ is also reapplied together with the $\mathbf{w}_t$ through a *heuristic reading-gate* $\mathbf{r}_t$ in DA-cell by Eqs. 5 and 6. The reading-gate's task is to update $\mathbf{d}_{t-1}$ such that if any slote-token e.g. SLOT_NAME, SLOT_AREA, DONT_CARE etc. appeared in the last step then the index of corresponding slot-value in $\mathbf{d}_{t-1}$ was set to zero with the help of Eq. 5:

$$\mathbf{d}_t = \mathbf{r}_t \odot \mathbf{d}_{t-1} \tag{5}$$

$$\mathbf{d}_t = tanh(\mathbf{W}_{rd}\mathbf{d}_t) \tag{6}$$

Due to the vanishing gradient problem in long sentences, improved version of LSTM is used as an s2s model and is called H-LSTM (Heuristically-gated-LSTM) as it takes heuristically modified $\mathbf{d}_t$ as input in each time-step:

$$\mathbf{i}_t = \sigma(\mathbf{W}_{wi}\mathbf{w}_t + \mathbf{W}_{hi}\mathbf{h}_{t-1} + \mathbf{W}_{di}\mathbf{d}_t) \tag{7}$$

$$\mathbf{f}_t = \sigma(\mathbf{W}_{wf}\mathbf{w}_t + \mathbf{W}_{hf}\mathbf{h}_{t-1} + \mathbf{W}_{df}\mathbf{d}_t) \tag{8}$$

$$\mathbf{o}_t = \sigma(\mathbf{W}_{wo}\mathbf{w}_t + \mathbf{W}_{ho}\mathbf{h}_{t-1} + \mathbf{W}_{do}\mathbf{d}_t) \tag{9}$$

$$\hat{\mathbf{c}}_t = \sigma(\mathbf{W}_{wc}\mathbf{w}_t + \mathbf{W}_{hc}\mathbf{h}_{t-1} + \mathbf{W}_{dc}\mathbf{d}_t) \tag{10}$$

$$\mathbf{c}_t = \mathbf{f}_t \odot \mathbf{c}_{t-1} + \mathbf{i}_t \odot \hat{\mathbf{c}}_t \tag{11}$$

$$\mathbf{h}_t = \mathbf{o}_t \odot tanh(\mathbf{c}_t) \tag{12}$$

where, $\mathbf{W}_{*,*}$ represents training weights and $\mathbf{i}_t$, $\mathbf{f}_t$, $\mathbf{o}_t$ are the input, forget and output gates of the LSTM-cell and $\hat{\mathbf{c}}_t$, $\mathbf{c}_t$ denotes step-wise local and global vector of the memory-cell. $\odot$ performs the element-wise multiplication.

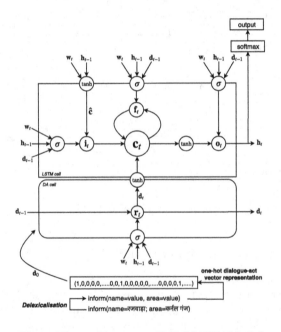

Fig. 2. The architecture of MSC-LSTM Model.

2.2   Modified-Semantically-Controlled Model (MSC-LSTM)

As we see, H-LSTM perfectly models the delexicalised data and incorporates DA details accurately up to a level in the generation. But this simple *content-planning* ability does not make it capable of handling the binary slots and the slots assigned with 'DONT_CARE' value, which can not be delexicalised. It is evident that the direct one-to-one matching of slot-value pairs and the corresponding surface-form realisation is not possible in H-LSTM. To address this problem, a mechanism for reading-gate $\mathbf{r}_t$ in the DA-cell, is proposed by Wen et al. [12], which remembers the associated phrases corresponding to slot-value pairs stochastically.

$$\mathbf{r}_t = \sigma(\mathbf{W}_{wr}\mathbf{w}_t + \mathbf{W}_{hr}\mathbf{h}_{t-1} + \mathbf{W}_{dr}\mathbf{d}_{t-1}) \tag{13}$$

$$\mathbf{d}_t = \mathbf{r}_t \odot \mathbf{d}_{t-1} \tag{14}$$

where, $\mathbf{W}_{wr}$, $\mathbf{W}_{hr}$ and $\mathbf{W}_{dr}$ are again the weight-matrix to learn the sequential pattern of key-phrases associated with the slot-value pairs. Thus, the updated reading-gate, if replaced in Eq. 5, would make the model more resilient to learn *delexicalised-phrases*.

We have proposed a modified version of SC-LSTM (MSC-LSTM) that includes the influence $\mathbf{d}_t$ not only in estimating the memory-cell value but also in readjusting, the weights of input, forget and output gates. MSC-LSTM shows better performance than SC-LSTM and H-LSTM as discussed in the result Sect. 4. Built on a typical LSTM architecture, MSC-LSTM has a memory-cell as well, see Fig. 2. The model updates the hidden layer as follows:

$$\mathbf{i}_t = \sigma(\mathbf{W}_{wi}\mathbf{w}_t + \mathbf{W}_{hi}\mathbf{h}_{t-1} + \mathbf{W}_{di}\mathbf{d}_{t-1}) \tag{15}$$

$$\mathbf{f}_t = \sigma(\mathbf{W}_{wf}\mathbf{w}_t + \mathbf{W}_{hf}\mathbf{h}_{t-1} + \mathbf{W}_{df}\mathbf{d}_{t-1}) \tag{16}$$

$$\mathbf{o}_t = \sigma(\mathbf{W}_{wo}\mathbf{w}_t + \mathbf{W}_{ho}\mathbf{h}_{t-1} + \mathbf{W}_{do}\mathbf{d}_{t-1}) \tag{17}$$

$$\hat{\mathbf{c}}_t = tanh(\mathbf{W}_{wc}\mathbf{w}_t + \mathbf{W}_{hc}\mathbf{h}_{t-1}) \tag{18}$$

$$\mathbf{c}_t = \mathbf{f}_t \odot \mathbf{c}_{t-1} + \mathbf{i}_t \odot \hat{\mathbf{c}} + tanh(\mathbf{W}_{dh}\mathbf{d}_t) \tag{19}$$

$$\mathbf{h}_t = \mathbf{o}_t \odot tanh(\mathbf{c}_t) \tag{20}$$

2.3   Attention-Based Encoder-Decoder (ENC-DEC)

This model is based on the architecture proposed earlier for Neural Machine Translation [19] which takes attention from all the slot-value during the encoding such that the sum of *attention-distribution* is equal to one as in Fig. 3. The $\mathbf{i}^{th}$ slot-value pair $\mathbf{z}_i$ is represented by the sum of distributed vectors of slot-embeddings $\mathbf{s}_i$ and value-embeddings $\mathbf{v}_i$.

$$\mathbf{z}_i = \mathbf{s}_i + \mathbf{v}_i \tag{21}$$

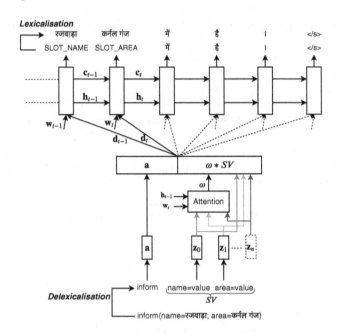

Fig. 3. The architecture of Attention-Based ENC-DEC model.

where, the size of vectors $\mathbf{s}_i$ and $\mathbf{v}_i$ is set equal to hidden-vector size. For adding semantics to $\mathbf{d}_t$, the slots and values are separately represented as a parameterized-vector [11]. At each time-step t, output of the DA-cell $\mathbf{d}_t$ is estimated by $\mathbf{z}_i$, weighted to its attentions.

$$\mathbf{d_t} = \mathbf{a} \oplus \sum_i \omega_{t,i}\mathbf{z_i} \qquad (22)$$

where $\mathbf{a}$ is the vector representation of DA-type and $\oplus$ is the sign of concatenation, and $\omega_{t,i}$ is the attention-weight of ith slot-value pair. At each iteration, the attention of all the slot-value pairs are calculated and normalized between 0 and 1 by Eqs. 23 and 24 respectively:

$$\beta_{t,i} = \mathbf{q}^T \tanh(\mathbf{W}_{hm}\mathbf{h}_{t-1} + \mathbf{W}_{mm}\mathbf{z}_i) \qquad (23)$$

$$\omega_{t,i} = \frac{\exp(\beta_{t,i})}{\sum_i \exp(\beta_{t,i})} \qquad (24)$$

In the decoding phase, we use a s2s-cell of H-LSTM model in the upper cell to estimate hidden vector $\mathbf{h}_t$ for predicting the next token.

3   Experiments

A simple word2vec[4] approach with *skip-gram* model is used to learn the word-embeddings for each token. Each word-embedding dimension is set to

---

[4] https://radimrehurek.com/gensim/index.html.

100 empirically. We have compared the models on two standard evaluation metrics: BLEU-score and error-rates (T-Error, S-Error). In general, T-Error denotes all miss-matches of slot-values in a DA and the corresponding generated utterance, while S-Error considers miss-matches only for those slots which do not have binary-values. Binary-value error denotes miss-matches of those slots which have binary-values. These models are compared with the standard baselines models listed below:

- Rule-based (HDC) generator,
- K-Nearest neighbour (KNN) based generator,
- N-Gram for class-based generator proposed by [6].

The dataset corpus is collected in the domain of searching a restaurant in Allahabad, a city in India by a group of pre-trained Hindi-speaking people. We managed to collect the corpus of **3K** pairs of DAs and corresponding utterances.

The RNNLG framework based generators were implemented using the PyDial Framework[5] in Theano library (PY2). The models are trained on an individual corpus partitioned in the ratio of 3:1:1 of training, validation and testing set with stochastic gradient-descent and back-propagation. L2-regularization is applied in order to prevent over-fitting with regularization factor 10^{-7}. Additionally, early stopping criteria based on the validation-error has also been incorporated to avoid the over-fitting.

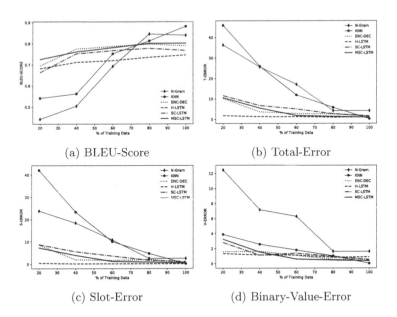

(a) BLEU-Score

(b) Total-Error

(c) Slot-Error

(d) Binary-Value-Error

Fig. 4. Comparison of RNNLG models on (%) of training-data.

---

[5] The Cambridge University Python Multi-domain Statistical Dialogue System Toolkit http://www.camdial.org/pydial/.

In the decoding phase, the model generates 20 utterances based on the beam-search decoding from which top 5 are selected based on *slot-error*[6][20]. The results depict the top performance of the models as obtained in the experiments.

4   Results and Analysis

In this section, we compare the output of all the models discussed in Sect. 2. The results are shown in Table 1. The HDC baseline model generates error-free utterances in terms of slot-value pairs but with lowest BLUE-score. The reason is that HDC is designed based on the pre-defined rules to generate rigid utterances which are different from the human collected utterances. Another drawback of the HDC model is scalability, which makes this model difficult to expand for the large domains. Next, the N-Gram baseline model shows improvement on BLEU-score compared to HDC but render worst S-Error due to missing slot-value pairs in the output. The third baseline model, KNN, which is based on the similarity of dialogue-acts of testing data to training and validating part of the corpus, shows the best results in terms of BLEU-score and Error-values, if 100% data is opted for training, testing and validation in the ratio of (3:1:1). But, if the training data size is reduced, its BLEU and Error-values get worse faster than the other models as shown in Fig. 4.

Table 1. Result of various models. (Underlined-models are the benchmarks. Errors are in percentage (%).)

Models	Test			Validation		
	BLUE	T-Error	S-Error	BLUE	T-Error	S-Error
HDC	0.26	0.00	0.00	-	-	-
N-Gram	0.85	4.20	2.57	-	-	-
KNN	0.88	0.28	0.24	-	-	-
ENC-DEC	0.79	1.43	0.83	0.77	3.01	2.06
H-LSTM	0.75	1.09	**0.19**	0.76	1.35	**0.39**
SC-LSTM	0.77	1.68	1.16	0.77	1.65	0.83
MSC-LSTM	**0.80**	**0.98**	0.59	**0.79**	**1.16**	0.77

The RNNLG models have shown greater performance than the above models in terms of scalability and adaptability. However, ENC-DEC result is not up to the mark as compared to H-LSTM, SC-LSTM and MSC-LSTM. This is due to the inherent limitation of attention-mechanism which does not prevent the slot repetitions in the generation process as shown in Fig. 5. This limitation is overcome in the later models by checking the slot repetition as shown in Fig. 4b.

---

[6] Utterances having minimum slot-error (S-Error) are selected.

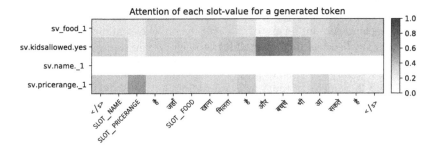

Fig. 5. ENC-DEC: key-phrase processing sequence.

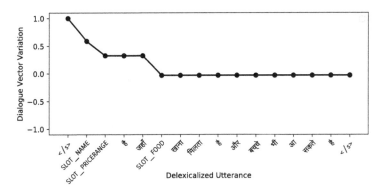

Fig. 6. H-LSTM: key-phrase processing sequence.

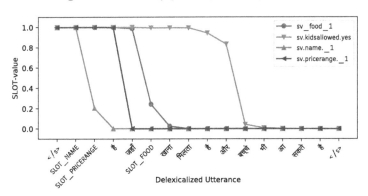

Fig. 7. MSC-LSTM: key-phrase detector.

All the Figs. 5, 6 and 7 are constructed corresponding to a DA[7] which represents how various models process the given DA.

---

[7] Ex.: inform(name=सतकार रेस्तरां;pricerange=महंगा;kidsallowed=yes;food= एशियन ओरि-एंटल).

While the heuristically-gated model has an advantage over semantically-controlled mechanism in terms of S-Error when the sentences are fully and properly delexicalised as in Fig. 4c. This is because the semantically-controlled

Table 2. Samples of top 5 realisations from all the RNNLG-models.

1.	inform(name=पिंड बलूची;area=सिविल लाइंस;pricerange=मध्यम;kidsallowed=no)
#	MSC-LSTM
	पिंड बलूची जो सिविल लाइंस में मध्यम एवं अच्छा भोजनालय है एवं यहाँ बच्चों को आने की अनुमति नहीं है।
	पिंड बलूची जो सिविल लाइंस में मध्यम एवं अच्छा भोजनालय है जहाँ बच्चों का जाना मना है।
	पिंड बलूची जो सिविल लाइंस में मध्यम एवं अच्छा रेस्टोरेंट है एवं यहाँ बच्चों को आने की अनुमति नहीं है।
	पिंड बलूची जो सिविल लाइंस में मध्यम एवं अच्छा भोजनालय है जहाँ बच्चों को आने की अनुमति नहीं है।
	पिंड बलूची जो सिविल लाइंस में मध्यम एवं अच्छा रेस्टोरेंट है जहाँ बच्चों का जाना मना है।
#	SC-LSTM
	पिंड बलूची सिविल लाइंस में मध्यम भोजनालय का रेस्टोरेंट है जहाँ बच्चों को आने की अनुमति नहीं है।
	पिंड बलूची सिविल लाइंस में मध्यम भोजनालय का भोजनालय है जहाँ बच्चों को आने की अनुमति नहीं है।
	पिंड बलूची सिविल लाइंस में मध्यम भोजनालय का रेस्टोरेंट है जहाँ बच्चों को प्रवेश की अनुमति नहीं है।
	पिंड बलूची सिविल लाइंस में मध्यम भोजनालय का भोजनालय है जहाँ बच्चों को प्रवेश की अनुमति नहीं है।
	पिंड बलूची सिविल लाइंस में मध्यम भोजनालय का भोजनालय है जहाँ बच्चे नहीं जा सकते हैं।
#	H-LSTM
	पिंड बलूची जो सिविल लाइंस में है , यहाँ मध्यम खाना मिलता है एवं वहाँ बच्चों को आने की अनुमति नहीं है।
	पिंड बलूची जो सिविल लाइंस में है , यहाँ मध्यम खाना मिलता है एवं यहाँ बच्चों को आने की अनुमति नहीं है।
	पिंड बलूची जो सिविल लाइंस में है , यहाँ मध्यम खाना मिलता है परन्तु वहाँ बच्चों को आने की अनुमति नहीं है।
	पिंड बलूची जो सिविल लाइंस में है , यहाँ मध्यम खाना मिलता है।
	पिंड बलूची जो सिविल लाइंस में है , यहाँ मध्यम खाना मिलता है एवं वहाँ बच्चों का जाना मना है।
#	ENC-DEC
	पिंड बलूची सिविल लाइंस में स्थित एक मध्यम भोजनालय है जहाँ बच्चों का जाना मना नहीं है।
	पिंड बलूची सिविल लाइंस में स्थित एक मध्यम भोजनालय है जहाँ बच्चों का आना मना नहीं है।
	पिंड बलूची सिविल लाइंस में स्थित एक मध्यम भोजनालय है जहाँ बच्चों का जाना मना है।
	पिंड बलूची सिविल लाइंस में स्थित एक मध्यम भोजनालय है जहाँ बच्चों को आने की अनुमति नहीं है।
	पिंड बलूची सिविल लाइंस में स्थित एक मध्यम भोजनालय है जहाँ बच्चों के लिए अनुमति नहीं है।
2.	inform(name=रजवाड़ा;area=कर्नल गंज;food=इंटरनेशनल;kidsallowed=yes)
#	MSC-LSTM
	रजवाड़ा कर्नल गंज में स्थित है जहाँ इंटरनेशनल खाना मिलता है और बच्चों को प्रवेश की अनुमति है।
	रजवाड़ा कर्नल गंज में स्थित है जहाँ इंटरनेशनल खाना मिलता है और बच्चे भी जा सकते हैं।
	रजवाड़ा कर्नल गंज में स्थित है , इंटरनेशनल खाना मिलता है और बच्चों को प्रवेश की अनुमति है।
	रजवाड़ा कर्नल गंज में स्थित है , इंटरनेशनल खाना मिलता है और बच्चे भी जा सकते हैं।
	रजवाड़ा कर्नल गंज में स्थित है जहाँ इंटरनेशनल खाना मिलता है और बच्चे बच्चे जा सकते हैं।
#	SC-LSTM
	रजवाड़ा कर्नल गंज में स्थित है जहाँ इंटरनेशनल खाना मिलता है और बच्चे भी जा सकते हैं।
	रजवाड़ा कर्नल गंज में स्थित है , में इंटरनेशनल खाना मिलता है और बच्चे भी जा सकते हैं।
	रजवाड़ा कर्नल गंज में स्थित है जहाँ इंटरनेशनल खाना मिलता है और वहाँ बच्चे भी जा सकते हैं।
	रजवाड़ा कर्नल गंज में स्थित है , में इंटरनेशनल खाना मिलता है और वहाँ बच्चे भी आ सकते हैं।
	रजवाड़ा कर्नल गंज में स्थित है , में इंटरनेशनल खाना मिलता है और बच्चे भी आ सकते हैं।
#	H-LSTM
	कर्नल गंज क्षेत्र में रजवाड़ा नाम का एक भोजनालय है जहाँ इंटरनेशनल खाना मिलता है और बच्चों को प्रवेश की अनुमति है।
	कर्नल गंज क्षेत्र में रजवाड़ा नाम का एक भोजनालय है जहाँ इंटरनेशनल खाना मिलता है और बच्चों का आना मना नहीं है।
	कर्नल गंज क्षेत्र में रजवाड़ा नाम का एक भोजनालय है जहाँ इंटरनेशनल खाना मिलता है और बच्चों को यहां की अनुमति है।
	कर्नल गंज क्षेत्र में रजवाड़ा नाम का एक भोजनालय है जहाँ इंटरनेशनल खाना मिलता है और बच्चों को आने की अनुमति है।
	कर्नल गंज क्षेत्र में रजवाड़ा नाम का एक भोजनालय है जहाँ इंटरनेशनल खाना मिलता है और बच्चों को प्रवेश की अनुमति है।
#	ENC-DEC
	रजवाड़ा कर्नल गंज में स्थित है जहाँ इंटरनेशनल खाना मिलता है और बच्चों को प्रवेश की अनुमति है।
	रजवाड़ा कर्नल गंज में स्थित है जहाँ इंटरनेशनल खाना मिलता है और बच्चे भी जा सकते हैं।
	रजवाड़ा कर्नल गंज में स्थित है , यहाँ इंटरनेशनल भोजन मिलता है। यहाँ पर बच्चों का आना मना नहीं है। यहाँ पर बच्चों का आना मना नहीं है।
	रजवाड़ा कर्नल गंज में स्थित है , यहाँ इंटरनेशनल भोजन मिलता है। यहाँ पर बच्चों का आना मना नहीं है। यहाँ पर बच्चों का आना मना है।
	रजवाड़ा कर्नल गंज में स्थित है जहाँ इंटरनेशनल खाना मिलता है और वहाँ बच्चे जा सकते हैं।

models work as a key-phrase detector as they map the slot-value as key to its corresponding generated output-phrase with the help of an additional reading-gate (Fig. 7). In contrast, the heuristically-gated model aligns the input DA to its associated phrase only for the slot-value pair which are delexicalised that makes it unsuitable for binary-value slot-value as shown in Fig. 4d.

The MSC-LSTM delivers better performance than the SC-LSTM as it incorporates the DA-vector $\mathbf{d}_t$ not only in the reading-gate but also in input, forget and output gates during the training, see its model architecture in Fig. 2. Example dialogue acts and their top-5 realisations for all the models are shown in Table 2.

5   Conclusion

In this paper, RNNLG framework has been adapted to explore and construct various RNN models for generating responses for a Hindi dialogue system. A general architecture in RNNLG framework is a combined process of sentence-planning and surface-realisation by a recurrent structure. Sentence planning has been investigated by three architectures of different capabilities: based on the gating mechanism (a) H-LSTM, (b) MSC-LSTM and (c) ENC-DEC. Several existing baselines e.g. HDC, N-Gram and KNN are used for comparison. Where, HDC and KNN models generate rigid utterances while class-based N-Gram and RNNLG based models have the ability to generate novel utterances based on the probability distribution of tokens in the training data. In terms of BLUE-score, T-Error and S-Error, RNNLG based models have shown better performances. MSC-LSTM is the best architecture among all the models due to its ability to remember key-phrases corresponding to DA-type and slot-value pairs through incorporating DA in a different way in the SC-LSTM architecture. In contrast, H-LSTM does not perform well for the binary slot-value pairs and ENC-DEC model allows the repetition of slot-value information.

Several limitations have been observed in the experiments, for future exploration. First one, if we take a DA which has two or more slot-values corresponding to one slot then its generated delexicalised utterance can not be lexicalised due to ambiguity in the order of slots-tokens. The second limitation is its dependence on the distribution of DA-structure in the corpus, hence we need to take care about combinations of DA and their occurrences during the data-collection. Third one is related to correct lexicalisation and delexicalisation of the slot-value pairs in a DA, which should be matched exactly with the collected utterances, hence we have to take care of DA-variance in the data during collection. And the fourth limitation is regarding the evaluation of generated utterances, we are able to check only on the basis of syntactic-error which is not sufficient as the generated utterances may show different meanings.

References

1. Walker, M.A., Rambow, O.C., Rogati, M.: Training a sentence planner for spoken dialogue using boosting. Comput. Speech Lang. **16**(3), 409–433 (2002). Spoken Language Generation
2. Stent, A., Prasad, R., Walker, M.: Trainable sentence planning for complex information presentation in spoken dialog systems. In: Proceedings of the 42nd Annual Meeting on Association for Computational Linguistics, ACL 2004. Association for Computational Linguistics, Stroudsburg, PA, USA (2004)
3. Malviya, S., Tiwary, U.S.: Knowledge based summarization and document generation using bayesian network. Procedia Comput. Sci. **89**, 333–340 (2016)
4. Langkilde, I., Knight, K.: Generation that exploits corpus-based statistical knowledge. In: Proceedings of the 36th Annual Meeting of the Association for Computational Linguistics and 17th International Conference on Computational Linguistics, ACL 1998/COLING 1998, Stroudsburg, PA, USA, pp. 704–710 (1998)
5. Rieser, V., Lemon, O.: Natural language generation as planning under uncertainty for spoken dialogue systems. In: Proceedings of the 12th Conference of the European Chapter of the ACL (EACL 2009), pp. 683–691. Association for Computational Linguistics, Athens, Greece, March 2009
6. Oh, A.H., Rudnicky, A.I.: Stochastic language generation for spoken dialogue systems. In: ANLP-NAACL 2000 Workshop: Conversational Systems (2000)
7. Mairesse, F., et al.: Phrase-based statistical language generation using graphical models and active learning. In: Proceedings of the 48th Annual Meeting of the Association for Computational Linguistics, pp. 1552–1561 (2010)
8. Mairesse, F., Young, S.: Stochastic language generation in dialogue using factored language models. Comput. Linguist. **40**(4), 763–799 (2014)
9. Konstas, I., Lapata, M.: A global model for concept-to-text generation. J. Artif. Intell. Res. **48**, 305–346 (2013)
10. Dušek, O., Jurčíček, F.: Sequence-to-sequence generation for spoken dialogue via deep syntax trees and strings. In: Proceedings of the 54th Annual Meeting of the Association for Computational Linguistics (Volume 2: Short Papers). pp. 45–51. Berlin, Germany (Aug 2016)
11. Mei, H., Bansal, M., Walter, M.R.: What to talk about and how? selective generation using LSTMs with coarse-to-fine alignment. In: Proceedings of the 2016 Conference of the North American Chapter of the Association for Computational Linguistics, San Diego, California, pp. 720–730 (2016)
12. Wen, T.H., Gašić, M., Mrkšić, N., Su, P.H., Vandyke, D., Young, S.: Semantically conditioned LSTM-based natural language generation for spoken dialogue systems. In: Proceedings of the 2015 Conference on Empirical Methods in Natural Language Processing, Lisbon, Portugal, pp. 1711–1721, September 2015
13. Wen, T.H., Young, S.: Recurrent neural network language generation for spoken dialogue systems. Comput. Speech Lang. **63**, 101017 (2020)
14. Jain, S., Malviya, S., Mishra, R., Tiwary, U.S.: Sentiment analysis: An empirical comparative study of various machine learning approaches. In: Proceedings of the 14th International Conference on Natural Language Processing (ICON-2017), pp. 112–121. NLP Association of India, Kolkata, India, December 2017
15. Jurčíček, F., Dušek, O., Plátek, O., Žilka, L.: Alex: a statistical dialogue systems framework. In: Sojka, P., Horák, A., Kopeček, I., Pala, K. (eds.) TSD 2014. LNCS (LNAI), vol. 8655, pp. 587–594. Springer, Cham (2014). https://doi.org/10.1007/978-3-319-10816-2_71

16. Hu, Z., Halberg, G., Jimenez, C.R., Walker, M.A.: Entrainment in pedestrian direction giving: how many kinds of entrainment? In: Rudnicky, A., Raux, A., Lane, I., Misu, T. (eds.) Situated Dialog in Speech-Based Human-Computer Interaction. SCT, pp. 151–164. Springer, Cham (2016). https://doi.org/10.1007/978-3-319-21834-2_14

17. Mikolov, T., Karafiát, M., Burget, L., Černocký, J., Khudanpur, S.: Recurrent neural network based language model. In: Eleventh Annual Conference of the International Speech Communication Association (2010)

18. Dhariya, O., Malviya, S., Tiwary, U.S.: A hybrid approach for hindi-english machine translation. In: 2017 International Conference on Information Networking (ICOIN), pp. 389–394. IEEE (2017)

19. Bahdanau, D., Cho, K., Bengio, Y.: Neural machine translation by jointly learning to align and translate. In: Proceedings of the ICLR (2015)

20. Wu, Y., et al.: Google's neural machine translation system: Bridging the gap between human and machine translation. arXiv preprint arXiv:1609.08144 (2016)

Bengali Handwritten Character Classification Using Transfer Learning on Deep Convolutional Network

Swagato Chatterjee[1], Rwik Kumar Dutta[1], Debayan Ganguly[2],
Kingshuk Chatterjee[3], and Sudipta Roy[4(✉)]

[1] BP Poddar Institute of Management and Technology, Kolkata, India
swagato.31dec@gmail.com, rwikdutta@outlook.com
[2] Government College of Engineering and Leather Technology, Kolkata, India
debayan3737@gmail.com
[3] Government College of Engineering and Ceramic Technology, Kolkata, India
kingshukchatterjee@gmail.com
[4] RCIL, Washington University in Saint Louis, St. Louis, MO 63110, USA
sudiptaroy1@yahoo.com

Abstract. Bengali is the sixth most popular spoken language in the world. Computerized detection of handwritten Bengali (Bangla Lekha) character is very difficult due to the diversity and veracity of characters. In this paper, we have proposed a modified state-of-the-art deep learning to tackle the problem of Bengali handwritten character recognition. This method used the lesser number of iterations to train than other comparable methods. The transfer learning on Resnet-50 deep convolutional neural network model is used on pretrained ImageNet dataset. One cycle policy is modified with varying the input image sizes to ensure faster training. Proposed method executed on BanglaLekha-Isolated dataset for evaluation that consists of 84 classes (50 Basic, 10 Numerals and 24 Compound Characters). We have achieved 97.12% accuracy in just 47 epochs. Proposed method gives very good results in terms of epoch and accuracy compare to other recent methods by considering number of classes. Without ensembling, proposed solution achieves state-of-the-art result and shows the effectiveness of ResNet-50 for classification of Bangla HCR.

Keywords: Bengali character · Classification · Transfer learning · Deep learning · Convolutional network

1 Introduction

Bengali is the second most widely spoken language in the Indian Subcontinent. More than 200 million people all over the world speaks this language and it is the sixth most popular language in the world. Thus proper recognition of handwritten Bengali characters is an important problem that has many noble applications

U. S. Tiwary and S. Chaudhury (Eds.): IHCI 2019, LNCS 11886, pp. 138–148, 2020.
https://doi.org/10.1007/978-3-030-44689-5_13

like Handwritten Character Recognition (HCR), Optical Character Recognition (OCR), and word recognition. However, Bengali letters are also more difficult to tackle than English. This is because apart from the basic set of characters-vowel and consonants, in Bengali script, there are conjunct-consonant characters as well which is formed by joining two or more basic characters. Many characters in Bengali resemble each other very closely, being differentiated only by a period or small line. Because of such morphological complexities and variance in the handwriting style, the performance of Bengali handwritten character recognition is comparatively quite lower than its English counterpart.

Ray and Chatterjee [21] did the first significant work in Bengali HCR. After that, many more researchers tried several other methods for improving the performance of Bangla HCR reported in [3,7]. Hasnat et al. [8], proposed an HCR capable of classifying both printed and handwritten characters by applying Discrete Cosine Transform (DCT) over the input image and Hidden Markov Model (HMM) for character classification. Wen et al. [30] proposed a Bangla numerals recognition method using principal component analysis and support vector machines. Liu and Suen [14], proposed a method of identifying both Farsi and Bangla Numerals. In Hassan and Khan [9], K-NN algorithm was used where features were extracted using local binary patterns. Das et al. [6], proposed a feature set representation for Bangla handwritten alphabets recognition which was a combination of 8 distance features, 24 shadow features, 84 quad tree based longest run features and 16 centroid features. Their accuracy was 85.40% on a 50 character class dataset. The above mentioned methods however used many handcrafted features extracted for small dataset which turned out to be unsuitable for deploying solutions.

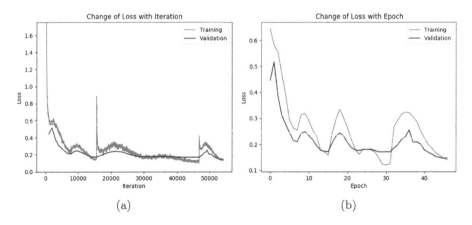

(a)   (b)

Fig. 1. Change of training and validation losses with each passing (a) iteration (b) epoch

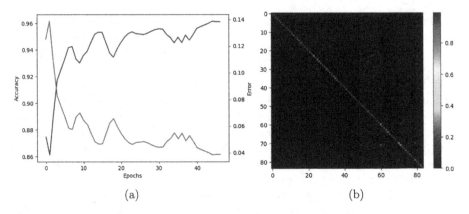

(a)   (b)

Fig. 2. (a) Change of accuracy and error rate with each passing epoch (b) Confusion matrix for all the 84 classes

With the advent of deep learning, it was possible to partly or fully eliminate the need for feature extraction. Recent methods [20,22] used Convolution Neural Networks (CNN) and improved the performance for Bangla character and digit recognition on a relatively large scale dataset. Sharif et al. [26], proposed a method on Bangla handwritten numeral classification where they bridged hand crafted feature using Histogram of Gradients (HOG) with CNN. Nibaran Das et al. [18], proposed a two pass soft-computing approach for Bangla HCR. The first pass combined the highly misclassified classes and provided finer treatment to them in the second pass. Sarkhel et al. [25], solved the problem using a multiobjective perspective where they trained a Support Vector Machine (SVM) classifier using the most informative regions of the characters. Alif et al. [1], used a modified version of ResNet-18 architecture with extra Dropout layers for recognizing Bangla Handwritten datasets on 84 classes; whereas Purkaystha et al. [19] used their own 7-layer CNN but on 80 classes. Alom et al. [2], used different CNN architectures on Bengali numerals, Characters and Special characters separately and reported DenseNet to be the best performing architecture. Majid and Smith [17], claimed to beat Alif et al. [1] but just presented their solution on 50 classes instead of 84 classes. Using auxiliary classifiers Saha and Saha [24], reported 97.5% accuracy, the procedure depended on ensemble learning. Individually, their accuracy was 95.67% and 92.43%. Although the more recent methods achieved better performance than the earlier methods, there's still scope for improvement in performance.

CNNs have been used quite successfully for similar problems in other languages as well. Ciresan and Meier [5] applied multi-column CNNs to recognize digits, alpha-numerals, Chinese characters, traffic signs, and object images. They surpassed previous best performances on many public databases, including MNIST digit image database, NIST SD19 alphanumeric character image dataset, and CASIA Chinese character image datasets. Korean or Hangul Handwritten character recognition system has also been proposed using Deep CNN [13].

The CNN Model proposed is ResNet-50 pre-trained with ILSVRC dataset [23] and fine-tuned using Transfer Learning. The objective of the paper is to use current best practices to train the model effectively and in as few iterations as possible. For better performance, optimization techniques like a slightly modified version of One Cycle Policy was used. In just 47 epochs, we were able to achieve 96.12% accuracy on BanglaLekha-Isolated dataset. The dataset had 84 classes, each for every character in the BanglaLekha-Isolated Dataset.

The CNN Model proposed is ResNet-50 pre-trained with ILSVRC dataset [23] and fine-tuned using Transfer Learning. The objective of the paper is to use current best practices to train the model effectively and in as few iterations as possible. For better performance, optimization techniques like a slightly modified version of One Cycle Policy was used. In just 47 epochs, we were able to achieve 96.12% accuracy on BanglaLekha-Isolated dataset. The dataset had 84 classes, each for every character in the BanglaLekha-Isolated Dataset.

Although a lot of previous work has been done on this topic as detailed in Sect. 1, improvements can still be achieved using recent advancements both in deep learning and also in model training procedures (like hyperparameter tuning as in Sect. 2.4, data augmentation, transfer learning etc.). The BanglaLekha-Isolated Dataset as described in Sect. 2.5 was used for evaluation because of its large sample size, inferior data quality (when compared to other available datasets which suits our purpose since it helps us better generalize) and large variance; and also because the output classes are balanced. The dataset consisted of 84 characters, which consisted of 10 Numerals, 50 basic characters and 24 frequently used compound characters. Section 3 describes about the results with discussion and finally we conclude our method in Sect. 4.

2   Methodology

In this section we discussed about the architecture and regularizations used to improve the performance.

2.1   Architecture

The model is a pre-trained Deep Convolutional Neural Network (CNN) called ResNet-50 [10]. We chose ResNet because of its heavy reliance on Batch Normalization and Dropout. These two techniques produces a regularizing effect on the model and prevents it from overfitting. Moreover, the identity-mappings or skip-connections in Resnets [10] help us tackle the vanishing gradient problem which in turn helps us train a deeper model which would be able give us better performance. The model has been pre-trained with the weights obtained by training the model with Imagenet Large Scale Visual Recognition Challenge (ILSVRC) [23] dataset. Thus, the initial model accepted images of size 224 * 224 px and classified the images to 1000 categories. We employed transfer learning to use this pre-trained model and modified it to classify 84 classes instead of 1000 by removing the Fully Connected (FC) layers of the original model and substituting

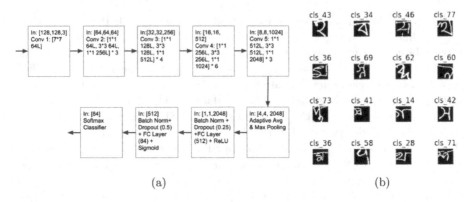

Fig. 3. (a) Architecture diagram of the model (b) Sample input and corresponding model output

them with new layers as given in Fig. 3a. We then used a softmax layer to give us the probability for each of the 84 classes and used Cross Entropy Loss as the loss function. The metric used for assessing the model performance is accuracy.

$$Accuracy = \frac{1}{N} \sum_{i=1}^{N} f(t_i, p_i) \tag{1}$$

where

$$f(t_i, p_i) = \begin{cases} 1 & p_i = t_i \\ 0 & p_i \neq t_i \end{cases} \tag{2}$$

and p_i and t_i is prediction and label for data point i respectively. We used AdamW Optimizer [16] and learning rate scheduler while training, which is a modified version of one cycle policy as has been explained in detail in Sect. 2.4.

2.2   Batch Normalization

Explosion and vanishing of gradients during backpropagation had been always been a problem. Hence to regularize the gradients Ioffe and Szegedy [12] came up with the solution of normalizing every minibatch after few layers. The algorithm picks up a mini-batch x from $\mathcal{B} = \{x_{1..m}\}$ and then apply the following equations:

$$\mu_{\mathcal{B}} \leftarrow \frac{1}{m} \sum_{i=1}^{m} x_i \tag{3}$$

$$\sigma_{\mathcal{B}}^2 \leftarrow \frac{1}{m} \sum_{i=1}^{m} (x_i - \mu_{\mathcal{B}})^2 \tag{4}$$

$$\widehat{x}_i \leftarrow \frac{x_i - \mu_{\mathcal{B}}}{\sqrt{\sigma_{\mathcal{B}}^2 + \epsilon}} \tag{5}$$

$$y_i \leftarrow \gamma \widehat{x}_i + \beta \equiv BN_{\gamma,\beta}(x_i) \tag{6}$$

Here γ and β are learnable parameters which help the backpropagation algorithm control the mean and variance of the batch and in turn help us use higher learning rate for training the model faster.

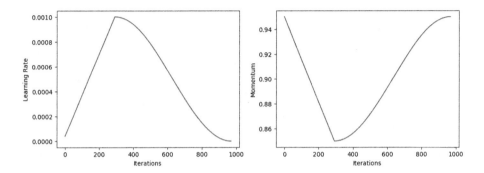

Fig. 4. Learning rate scheduling

2.3   Dropout

Dropout [29] is another regularization technique in which the activation of random neurons with a probability drawn from a Binomial or Gaussian distribution in a layer is clamped down to zero. This has a regularizing effect on the neurons and the backpropagation hence forces the neurons find new pathways. Hence dropout ensures that the network learns, and not memorizes, thus preventing overfitting

2.4   Learning Rate and Optimizers

For our experiments we used a modified version of One-cycle policy [28], a Learning Rate Scheduler, which it's authors also call superconvergence. In superconvergence the learning rate η goes up till the epoch T, using the current epoch T_i bounded by η_{max} and η_{min}. This gives a linear warm up at iteration t as given in Eq. 7.

$$\eta_t = \eta_{min} + (\eta_{max} - \eta_{min})\frac{T_i}{T} \tag{7}$$

After about 30% of the total iterations later, the learning rate goes down like half a cosine curve Eq. 8 called Cosine Annealing [15].

$$\eta_t = \eta_{min} + \frac{1}{2}(\eta_{max} - \eta_{min})(1 + \cos(\frac{T_i}{T}\pi)) \tag{8}$$

A sample of the learning rate scheduler used by us is shown in Fig. 4. This policy reduce the number of epochs required to train the model. Moreover, we employ a practice called Learning Rate Finding [27] in which the mini-batches of data is passed through the model and it's loss is measured against slowly increasing learning rate; from values as low as 10^{-7}; till the loss explodes. The learning rate chosen is the 10^{-1} of the minimum value of the loss vs learning rate curve. This learning rate is the η_{max}. The η_{min} is chosen to be $\frac{1}{25}\eta_{max}$.

One-cycle policy also utilizes a momentum scheduler, which unlike the learning rate first linearly goes down to a minimum value, the moment the learning rate scheduler enters its cosine-annealing phase, it follows the cosine curve to its initial value. For all our experiments we kept the range from 0.85 to 0.9. We use the AdamW [16] optimizer for optimization. AdamW employs a different strategy for updating the weights using L_2 weight decay parameter, λ.

2.5   BanglaLekha-Isolated Dataset

The BanglaLekha-Isolated dataset [4] was compiled from handwriting samples of both male and female native Bangla speakers of different states of Bangladesh, of age range 4 to 27. Moreover a small fraction of the samples belong from people with disabilities. Table 1 summarizes the dataset. This dataset doesn't have class imbalance, i.e. the number of images in each character class is almost equal.

Table 1. Class and image frequency in BanglaLekha-Isolated dataset

Character type	Classes	Counts
Basic character (Vowel + Consonants)	50	98,950
Numerals	10	19,748
Conjunct-consonant characters	24	47,407
Total	84	166,105

These images were then preprocessed by applying colour-inversion, noise removal and edge-thickening filters. Out of the 166,105 samples, we used 25% of the data (i.e. 41,526) for validation set and 75% of the data (i.e. 124,579) for training set. The image size of the dataset varied from 110 px * 110 px to 220 px * 220 px. Few samples from the dataset are shown in Fig. 3b.

2.6   Training Steps

We trained our model on a ResNet-50 CNN pretrained on ImageNet-Object Classification dataset with data augmentation of zoom, lighting and warping with mirror padding. We used 25% of the dataset for testing and the rest for training. The batch size was set to 128. At first, we scaled down the images to 128 px $\times$ 128 px and trained only the randomly initialized FC layers for 8 epochs. Then, all the layers were unfreeze for fine-tuning with discriminative layer training [11] for different layers. The earlier layers needs to be fine-tuned less and hence has a smaller learning rate than the later layers which needs to be fine-tuned more. We then repeat the above two steps after resizing the images to 224 × 224 px. Note that for 224 × 224 px, the batch size was reduced to 64. After this the bottom layers were frozen again and the model is finetuned with 128 × 128 px images. A detailed tabular view of the various steps are given in Table 2.

Table 2. Summary of training (Eps. = Epochs)

Eps	Img. size	Batch size	Learning rate	Frozen	Train. loss	Valid. loss	Accuracy %
8	128	128	$2 * 10^{-2}$	Yes	0.25	0.20	94.24
8	128	128	$(4 * 10^{-5}, 1 * 10^{-3})$	No	0.15	0.17	95.33
8	224	96	$1 * 10^{-2}$	Yes	0.16	0.17	95.37
8	224	96	$(8 * 10^{-6}, 2 * 10^{-4})$	No	0.12	0.17	95.56
15	128	256	$(3 * 10^{-4}, 3 * 10^{-2})$	Yes	0.14	0.14	96.12

3   Results and Discussion

The training and validation loss are plotted in Fig. 1 to ensure that the model is neither overfitting nor underfitting. The change of metrics is shown in Fig. 2a. The confusion matrix has been plotted in Fig. 2b. At the end, we were able to achieve an accuracy of 96.12% on the validation set. Table 3 compares our method with that of other researchers, and from the table considering number of classes and without using Ensemble Learning, the proposed solution achieves state of the art results in Isolated Handwritten Bengali Character Recognition. Majid and Smith [17]'s solution provides better accuracy but is based on only 50 classes. Although Saha and Saha [24]'s solution does achieve better result when using ensembling (using two CNNs of accuracies: 95.67% and 92.43%), our solution uses 47 epochs for training while their solution uses 500 epochs.

Table 3. Comparison of isolated handwritten Bengali character recognition techniques on BanglaLekha-isolated

Researcher	Classes	Method	Accuracy
Purkaystha et al. [19]	50	Vanilla CNN	89.01%
Alif et al. [1]	84	ResNet-18[a]	95.10%
Majid and Smith [17]	50	Non-CNN	96.80%
Saha and Saha [24]	84	Ensemble CNN[b]	97.21%
Proposed work	84	ResNet-50	96.12%

[a]Modified Resnet-18 with Dropout (0.2) after each block, validation set of size 20%
[b]Ensembles two CNN of accuracy 95.67% and 92.43%

After analysis of the misclassified examples, it was seen that there are quite a few data points whose ground truth is mislabeled. For the purpose of accurate benchmarking, we haven't removed those datapoints. But if those mislabeled datapoints are removed from the dataset, the accuracy will improve further. Our experimentation also contains some limitations. The Bengali language has more conjunct-consonant characters than just the 24 frequently used ones which are present in the BanglaLekha-Isolated dataset. This means that even though the character set present in BanglaLekha-Isolated dataset represents the major chunk of the entire Bengali corpus, it doesn't contain every single character. Hence, the performance presented is for those 84 characters only. However, to the best of our knowledge, there are no datasets which not only contains sufficient samples of all the characters in Bengali language but is also of good quality.

4 Conclusion

Our model was able to achieve an accuracy of 97.12% on the BanglaLekha-Isolated dataset. Without ensembling, proposed solution achieves state-of-the-art result and shows the effectiveness of ResNet-50 for classification of Bangla HCR. The best practice is used to train the model effectively and in as few iterations as possible. For better performance, optimization techniques like a slightly modified version of one cycle policy was used. In just 47 epochs, we were able to achieve 97.12% accuracy on standard datasets.

References

1. Alif, M.A.R., Ahmed, S., Hasan, M.A.: Isolated bangla handwritten character recognition with convolutional neural network. In: 2017 20th International Conference of Computer and Information Technology (ICCIT), pp. 1–6. IEEE, Dhaka, December 2017. http://ieeexplore.ieee.org/document/8281823/
2. Alom, M.Z., Sidike, P., Hasan, M., Taha, T.M., Asari, V.K.: Handwritten Bangla character recognition using the state-of-the-art deep convolutional neural networks. Comput. Intell. Neurosci. **2018**, 1–13 (2018). https://www.hindawi.com/journals/cin/2018/6747098/
3. Bhattacharya, U., Shridhar, M., Parui, S.K.: On recognition of handwritten bangla characters. In: Kalra, P.K., Peleg, S. (eds.) ICVGIP 2006. LNCS, vol. 4338, pp. 817–828. Springer, Heidelberg (2006). https://doi.org/10.1007/11949619_73
4. Biswas, M., et al.: BanglaLekha-Isolated: a multi-purpose comprehensive dataset of Handwritten Bangla Isolated characters. Data Brief **12**, 103–107 (2017)
5. Ciresan, D.: Multi-Column Deep Neural Networks for Offline Handwritten Chinese Character Classification, pp. 1–6. IEEE, Killarney (2015). http://ieeexplore.ieee.org/document/7280516/
6. Das, N., Basu, S., Sarkar, R., Kundu, M., Nasipuri, M., Basu, D.K.: An Improved Feature Descriptor for Recognition of Handwritten Bangla Alphabet. arXiv:1501.05497 [cs], January 2015
7. Forkan, A.R.M., Saha, S., Rahman, M.M., Sattar, M.A.: Recognition of conjunctive Bangla characters by artificial neural network. In: 2007 International Conference on Information and Communication Technology, ICICT 2007, pp. 96–99. IEEE (2007)

8. Hasnat, M.A., Habib, S.M., Khan, M.: A high performance domain specific OCR For Bangla script. In: Sobh, T., Elleithy, K., Mahmood, A., Karim, M.A. (eds.) Novel Algorithms and Techniques In Telecommunications. Automation and Industrial Electronics. Springer, Dordrecht (2008). https://doi.org/10.1007/978-1-4020-8737-0_31

9. Hassan, T., Khan, H.A.: Handwritten Bangla numeral recognition using local binary pattern. In: 2015 International Conference on Electrical Engineering and Information Communication Technology (ICEEICT), pp. 1–4. IEEE (2015)

10. He, K., Zhang, X., Ren, S., Sun, J.: Deep Residual Learning for Image Recognition. arXiv:1512.03385 [cs] , December 2015

11. Howard, J., Ruder, S.: Universal Language Model Fine-tuning for Text Classification. arXiv:1801.06146 [cs, stat], January 2018

12. Ioffe, S., Szegedy, C.: Batch Normalization: Accelerating Deep Network Training by Reducing Internal Covariate Shift. arXiv:1502.03167 [cs], February 2015

13. Kim, I.-J., Xie, X.: Handwritten Hangul recognition using deep convolutional neural networks. IJDAR **18**(1), 1–13 (2014). https://doi.org/10.1007/s10032-014-0229-4

14. Suen, C.Y.: A new benchmark on the recognition of handwritten bangla and Farsi numeral characters. Pattern Recogn. **42**(12), 3287–3295 (2009)

15. Loshchilov, I., Hutter, F.: SGDR: Stochastic Gradient Descent with Warm Restarts. arXiv:1608.03983 [cs, math], August 2016

16. Loshchilov, I., Hutter, F.: Fixing Weight Decay Regularization in Adam. arXiv:1711.05101 [cs, math], November 2017

17. Majid, N., Smith, E.H.B.: Introducing the Boise state Bangla handwriting dataset and an efficient offline recognizer of isolated Bangla characters. In: 2018 16th International Conference on Frontiers in Handwriting Recognition (ICFHR), pp. 380–385. IEEE (2018)

18. Das, N., Basu, S., Saha, P.K., Sarkar, R., Kundu, M., Nasipuri, M.: Handwritten Bangla character recognition using a soft computing paradigm embedded in two pass approach. Pattern Recogn. **48**(6), 2054–2071 (2015). https://doi.org/10.1016/j.patcog.2014.12.011

19. Purkaystha, B., Datta, T., Islam, M.S.: Bengali handwritten character recognition using deep convolutional neural network. In: 2017 20th International Conference of Computer and Information Technology (ICCIT), pp. 1–5. IEEE (2017)

20. Rahman, M.M., Akhand, M.A.H., Islam, S., Chandra Shill, P., Hafizur Rahman, M.M.: Bangla handwritten character recognition using convolutional neural network. Int. J. Image Graph. Signal Process. **7**(8), 42–49 (2015). http://www.mecs-press.org/ijigsp/ijigsp-v7-n8/v7n8-5.html

21. Ray, A., Chatterjee, B.: Design of a nearest neighbour classifier system for Bengali character recognition. IETE J. Res. **30**(6), 226–229 (1984). http://www.tandfonline.com/doi/full/10.1080/03772063.1984.11453273

22. Roy, S., Das, N., Kundu, M., Nasipuri, M.: Handwritten isolated Bangla compound character recognition: a new benchmark using a novel deep learning approach. Pattern Recogn. Lett. **90**, 15–21 (2017). https://linkinghub.elsevier.com/retrieve/pii/S0167865517300703

23. Russakovsky, O., et al.: ImageNet Large Scale Visual Recognition Challenge. arXiv:1409.0575 [cs], September 2014

24. Saha, S., Saha, N.: A lightning fast approach to classify Bangla handwritten characters and numerals using newly structured deep neural network. Procedia Comput. Sci. **132**, 1760–1770 (2018)

25. Sarkhel, R., Das, N., Saha, A.K., Nasipuri, M.: A multi-objective approach towards cost effective isolated handwritten Bangla character and digit recognition. Pattern Recogn. **58**, 172–189 (2016). https://linkinghub.elsevier.com/retrieve/pii/S0031320316300437
26. Sharif, S.M.A., Mohammed, N., Mansoor, N., Momen, S.: A hybrid deep model with HOG features for Bangla handwritten numeral classification. In: 2016 9th International Conference on Electrical and Computer Engineering (ICECE), pp. 463–466 (2016)
27. Smith, L.N.: Cyclical Learning Rates for Training Neural Networks. arXiv:1506.01186 [cs], June 2015
28. Smith, L.N.: A disciplined approach to neural network hyper-parameters: Part 1 - learning rate, batch size, momentum, and weight decay. arXiv:1803.09820 [cs, stat], March 2018
29. Srivastava, N., Hinton, G., Krizhevsky, A., Sutskever, I., Salakhutdinov, R.: A simple way to prevent neural networks from overfitting. J. Mach. Learn. Res. **15**, 1929–1958 (2014). http://jmlr.org/papers/v15/srivastava14a.html
30. Wen, Y., Lu, Y., Shi, P.: Handwritten Bangla numeral recognition system and its application to postal automation. Pattern Recogn. **40**(1), 99–107 (2007). https://doi.org/10.1016/j.patcog.2006.07.001

Vision Based Interactions

Predicting Body Size Using Mirror Selfies

Meet Sheth$^{(\boxtimes)}$ and Nisheeth Srivastava$^{(\boxtimes)}$

IIT Kanpur, Kanpur, Uttar Pradesh, India
mhsheth37@gmail.com, nsrivast@cse.iitk.ac.in

Abstract. Purchasing clothes that fit well on e-commerce portals can be problematic if consumers do not trust the fit of clothes based on specified size labels. This is especially a problem in developing countries, where size labels are not adequately standardized. In this paper, we introduce a system that can take a person's mirror selfie as input and accurately predict anthropometric measurements using that image. These anthropometric measurements can then be used to predict clothing fit based on supplier-specific measurement-label mappings that are available or can easily be developed by e-commerce clothing retailers. The key novelty of our proposal is our use of mirror selfies, which physically ensures that an object of standardized and known size, a cellphone, is present in an image at a predictable orientation and location with the person being photographed. For predicting measurements, we experimented with a number of regression models. Empirical testing showed that the best regression models yield $\leq 5\%$ test set error with respect to 11 tailor-derived body measurements for each of 70 male subjects. The empirical success of our proposal leads us to believe that our proposed approach may considerably simplify the task of online body size prediction.

Keywords: Personalization · Image processing · Machine learning · e-commerce

1 Introduction

While buying clothes in the real world, we visit brick and mortar stores and the tailor takes our measurements in standardized postures. Nowadays with the rapid increase in technology worldwide and with the growth of e-commerce, internet has been used for selling various products including apparel, locally and internationally. A crucial challenge in selling clothes online is finding a comfortable size for subject. Each brand has a different size label for the same body size. Due to this, people face difficulties in buying apparel online. So sellers have to budget for a high rate of replacements for this kind of product. Because of this, the online apparel and footwear sales are far behind others such as books and CDs, due to the uncertainties of correct sizing and difficulties with product evaluation (for example texture, color, finish) [3]. This situation strongly demands a system that can solve this problem somehow. Some work has been done to

© Springer Nature Switzerland AG 2020
U. S. Tiwary and S. Chaudhury (Eds.): IHCI 2019, LNCS 11886, pp. 151–162, 2020.
https://doi.org/10.1007/978-3-030-44689-5_14

solve this problem, as we briefly discuss below. In this paper, considering the customer's point of view, we present a simple and user-friendly way of predicting anthropometric measurements from a single photo.

2 Related Work

Majority of researchers who have done work of predicting measurements from an image, considering both front and side view of subject solve the problem using the following outline. The first thing they do is to generate silhouette from an image; then they extract feature points (or characteristic points) on that. Here feature points mean points that indicate different body part like front shoulder, side waist, elbow, etc. After this deformation technique like FFD [1] is applied on silhouette and existing 3-D human body template. Which will result in deformed 3-D human body template. This deformed human body template then used to predict all required measurements. Method of finding silhouette, method of extracting feature points and method of deforming 3-D template model make a difference in all research work that has been done.

But in the last decade, more focus has been given to a 3-D human scan rather than using an image for prediction. The reason behind this is low error rate compare to the image-based system. The general outline of solving a problem using a 3-D human scan is as follows. Using 3-D scanner 3-D scan of subject is generated. The first step is to clean the scan (i.e. removing noise), then segmentation has been done, which includes both separating subject's scan from the background and segmenting different body parts like chest, thigh, etc. After this feature points are extracted and then measurements are predicted using some machine learning technique or using some other method. But the setup for generating a 3-D scan using a scanner is very complex and expensive. And also it is not feasible for an application like online shopping of apparel.

In this section related work is mentioned in two parts. First part shows the work of predicting measurements from an image and in the second part it shows from a 3-D scan of subject.

2.1   2-D Image-Based Prediction System

2D image-based anthropometric measurement and clothing sizing system was mentioned in [2]. For data collection, specific setup was made which shown in Fig. 1. As we can see a total of 2 photographs was taken at the same time using two camera. The same camera was used for all the subjects. Blue backdrop embedded with calibration markers was used as shown in setup.

The steps involved for predicting measurement are pre-processing of both the views of image, calibration of camera, segmentation, landmark detection, prediction of measurements in given sequence. This work was done on 349 subjects and a total of 5 measurements were considered for prediction. The authors' claimed results show a ≤ 2 mm difference between actual and predicted values for

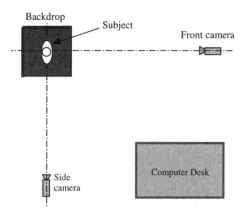

Fig. 1. Data collection setup

body measurements, but this error is calculated summed over the entire participant cohort, not for individuals. Thus, their results show accurate cohort level predictions rather than useable individual level predictions.

Another similar kind of approach was proposed by [3] with some variation in the modeling of measurement. In this work, the formula for measuring the circumference of neck and chest changed to ellipse and rectangular ellipse respectively unlike cylindrical given in [2].

Work of [4] is interesting due to its combined approach of using both image-based and example-based modeling technique to create human body model. A total of 21 subjects were considered for their study. Using body shape profile, of an already existing database of 5000 subjects on given two images, they have predicted body profile for a subject. For a reconstruction of 3-D shape feature, they have used ct-FFD [1] algorithm on a template model for deforming it according to subject's 2-D silhouette. Results by them show <2 cm mean absolute error for all measurements. The drawback of this approach is manual detection of feature points in the first phase of the algorithm, which is a tedious and time-consuming task if we use this system for online shopping.

Recent work of [5] tries to improve accuracy of the image-based system by adopting global and local deformation methods. Outline of their work is common as others work except this deformation techniques. From the 2-D image, silhouette extraction, model deformation, and matching skeleton systems are usual steps. Finally, an anthropometric database including 26 items of dimensions and a 3-D model of the subject can be outputted from their system. But some of their predicted measurements show a mean difference of more than 2 cm, which is not allowed for some specific measurements. The approach presented in this paper does not involve creating a 3-D human model which makes our prediction model much simpler.

2.2   Predicting Measurements from 3-D Human Scan

The performance of commercial 3-D scanning systems for predicting anthropometric measurements were shown in several studies.

Work of [6] presents an approach that is based on alignment for predicting anthropometric measurements from high-resolution 3-D human scan. They propose a solution which they called model-based anthropometry. For scanning the noise this model-based solution is robust. This brings the scan into registration with the registered body scans database. The registration process has been described in detail in [7]. After this, from the registered model they extract features. And finally mapping from these features to measurements were learned using regularized linear regression. Error matrix used by them was Average Mean Absolute Difference (AMAD) and its values for male and female are 10.09 mm and 10.02 mm respectively, which is higher than allowable error.

In [8], the authors test using a large dataset of 400 female subjects. In this work, they first extracted representative body size from the 3-D scan using factor analysis and garment knowledge. From these body sizes, those which are easy to measure were selected as feature parameters (FP). On these FPs, the combination of radial basis function and multiple linear regression were used to predict detailed measurements. Their result shows less than 6% average percentage error for all the prediction.

The work of [9] shows a relatively better result than others. Using a 3-D laser scanner, input scan of subject was made. This study was done considering only 16 subjects, out of which eight were male and the remaining eight were female. The approach used by them includes the creation of curves/surfaces, creation of different planes dimension, creation of splines and creation of points with a known distance in sequence. After this using CAD model, different body dimensions were predicted. A total of 5 measurements were included in the prediction. The absolute mean difference between actual and predicted measurements is <1 cm for all 5 measurements.

Recent work of [10] on 40 individuals(21 male + 19 female) is much more clear and detailed in comparison to other research work. The general processing path in this work includes body segmentation, Gaussian curvature calculation, characteristic points localization and then using those points total of 34 measurements were predicted. Median percentage error of all the measurements among 40 subjects lies between −5 to 0%. The negative sign in percentage error is because of the consideration of direct difference between actual and manual measurement in the formula of percentage error instead of taking absolute difference in the numerator.

Another much recent work by [11] represents a method for predicting measurements that are insensitive to body posture and shape in input 3-D scan. They proposed a novel modeling approach based on ensemble learning to extract focal features on 3D human body. The extraction of focal features was done via random forest regression analysis of geodesic distance. The predicted displacements of the randomly sampled points on the new 3-D human body surface are used to vote for the desired semantic focal feature. Twenty subjects took part in this study. A total of four measurements were considered in prediction. The experiment results show that the average error of shoulder width and girth(which includes bust, hip, and waist) are 1.332 cm and 0.7635 cm, respectively. The limitation of this work is its computational inefficiency.

The error rate in all work that is presented in Sect. 2.2 is less than the work which is based on an image. But considering its practical application like online shopping, having a specific hardware setup for 3-D laser scanning is expensive and complicated, therefore not feasible. This nullify the purpose of easy and smooth online shopping of apparel. Hence our work represents a very cost-effective approach for getting important measurements from the 2-D image of subject via mirror selfie.

3   Approach

Our goal is to develop a model that can predict anthropometric measurements from a mirror selfie taken by a subject. We divided this into two parts. The first was to collect a dataset of (mirror selfie + measurements) of different individuals and the second part is to create a model that can predict anthropometric measurements of subject. We collected a total of 11 measurements covering all parts of the subject's body. Four of them are linear measurements which include shoulder width, arm length, shirt length, pant length and other 7 are circumferential measurements which include chest circumference, waist, hips circumference, thigh, bottom of pant, wrist, collar of a shirt.

The second was to predict body size in these 11 dimensions given an image. An image can be seen in 3 parts: background, subject and cell phone of subject. Prediction task further can be divided into two sub parts. One is to detect the subject, and its cell phone and other is to predicting all 11 measurements.

3.1   Data Collection

For each individual mirror selfie, cell phone model name which is used to take a selfie and 11 measurements are required. Here mirror selfie is defined as an image/picture which is taken by subject in standing position in front of a mirror holding cell phone in one hand. Example mirror selfie is shown in the Fig. 2.

Fig. 2. Example mirror selfie

The importance of a cell phone model is to determine pixel to mm/cm scaling factor as there is no other calibration object available in the image. Required 11 measurements can be divided into linear and circumferential measurements. Linear measurements include shirt length, pant length, sleeve length, height and shoulder width from backside. Circumferential measurements include chest circumference, waist, hip circumference, thigh circumference, the bottom periphery of pant, neck circumference(collar) and wrist circumference. So for each subject mirror selfie, cell phone model and 11 body measurements are required. Total of 70 male subjects took part in an offline data collection procedure. The age of male subjects varies from 20 to 60 years. But the majority of them range between 20 to 30 years.

3.2 Prediction Model

The prediction task can also be considered as being composed of three parts. The first sub-task is object detection, second one is data preparation, and the third one is to predict measurements.

Object Detection: From a given mirror selfie, we need to locate a person and his cell phone. For this we used a pre-trained faster R-CNN model from the TensorFlow object detection API. This helps in locating a person and cell phone by drawing a bounding box around the subject and his cell phone. The input model takes an image and gives coordinates of bounding boxes for phone and person both as output.

Data Representation: We are using supervised learning for training the model. For this we need data in (X, y) form, where X = data, y = label. In our data we have measurements of subjects, so labels are directly available. But as a data point, we have an image for all the subjects. For X we need to generate a feature vector for each image. For this purpose, we developed a task-specific representation built using three ratios α, β, and γ obtained from the raw image, which are defined as follows (Fig. 3).

α = person pixel width/image pixel width
β = 155 - actual height of cell phone (mm)
γ = phone pixel width/image pixel width

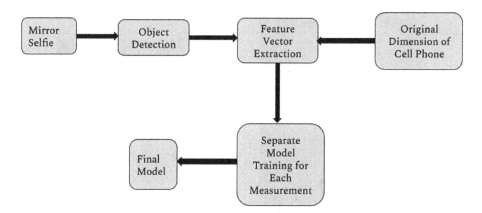

Fig. 3. Outline of training phase of prediction model

All three variables have been introduced by assuming that all measurements are proportional to cell phone height and our task is to find proportionality constant. This assumption gives a way to convert an image into some meaningful feature vector, which will be used to train a model later. Significance of each variable is mentioned below.

Alpha: It is defined as a ratio of person width and image width. Getting image width is easy because we have image resolution, so we know the width of an image in pixel. Object detection detects subject inside an image, so we have coordinates of the bounding box which bounds subject. Here the width of a bounding box is taken as width of a person. The purpose of the alpha parameter is to determine the proximity of subject from a mirror. If a subject is close to mirror, then he will cover more portion of the image compare to when he stands far from mirror. If we consider measurement shirt length, then this implies that the pixel length of shirt is more when subject is close to the mirror. For most of the data points, alpha lies between [0.4, 0.6]. Alpha = 0.4 means that subject is standing relatively at more distance from the mirror. And Alpha = 0.6 means subject is very close to the mirror. See below figure for example (Fig. 4).

Beta: It is defined as (155 - actual height of cell phone in mm). Here we are using machine learning techniques to train model, so there is no direct involvement of calibration object (cell phone). But in order to give an indirect effect of calibration object, we introduce variable beta. For the sake of understanding, let us assume that shirt length is proportional to phone height. The prediction model needs to find effective factor i.e. proportionality constant. And then we multiply it to the height of cell phone. But as we know that different cell phone is of different height. So we need to take it into consideration because we multiply phone's height to an effective constant (proportionality constant). For this, we take 155 mm as an average height of cell phone considering phone models nowadays. If the height of phone model is greater than 155 mm then beta will

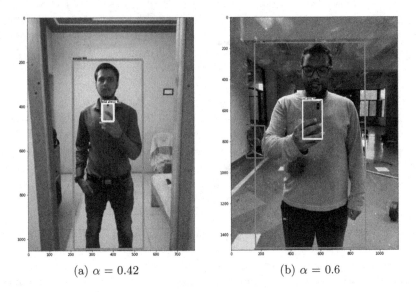

(a) $\alpha = 0.42$ (b) $\alpha = 0.6$

Fig. 4. Example images explaining proximity from mirror

be negative which says that effective factor should be less than its usual range. A similar argument can be made in case of positive beta.

Gamma: It is defined as a ratio of pixel width of phone model and pixel width of image. The pixel width of an image is easy to find because we have image resolution. The object detection module detects cell phone, so we have coordinates of the bounding box around the cell phone. Which gives us approximate width of phone model in pixel. The significance of this variable is to find the proximity of the phone to mirror. So while taking a selfie if the phone is very close to mirror then its pixel height is going to be more relatively, and due to this effective factor needs to be a bit smaller than the normal range.

Now considering supervised learning data point $(X, y)_i$ can be defined as

$$X_i = (\alpha, \beta, \gamma)$$
$$y_i = (actual\ measurement)$$

Model Training: We did separate training for all 11 measurements. So finally, we have 11 different models for each individual measurement. Total we have data of 70 male subjects and 71–29% train-test split was used to prepare the model. For model training, regression techniques like linear regression, support vector regression (SVR), k-nearest neighbor and decision tree regression were tried. The value of k is 5 in k-nearest neighbor. Decision tree was implemented using scikit-learn library of python, which uses optimised version of the CART (Classification and Regression Trees) algorithm. All models were trained using scikit-learn python library. Performance of each model by a different method is mentioned in the Result section.

Model Interpretation: Let us consider prediction of measurement shirt length. As mentioned earlier, that shirt length is proportional to cell phone height is taken as an assumption. In order to get better insight into the proposed parameters α, β, γ consider the equation given below.

$$SL = [4.5 \pm \frac{\alpha}{2} + \frac{\beta}{50} + 2.5 \times (0.1 - \gamma)] \times PH \tag{1}$$

where:
SL = shirt length, PH = cell phone height in pixel
Here, the first term is clearly a scaling factor defined using three variables α, β, γ derived from each individual's data. Let us consider the first term (4.5) in the proportionality constant, which tells us that shirt length is assumed 4.5 times to PH. But it can not be true in all cases. So α is introduced, which informs the model about the proximity of the person from the mirror. According to the value of α we add or subtract something from 4.5 and that makes the scaling factor more effective. Similarly other two term in proportionality constant can be explained. So overall, α, β, γ were introduced for making the scaling factor more effective by accounting for variability introduced by variation in peoples' posing for mirror selfies.

Notice that the effective scaling factor is linear in α, β, γ. So we can further rearrange the equation above slightly for better understanding.

$$\frac{SL}{PH} = 4.75 \pm \frac{\alpha}{2} + \frac{\beta}{50} - 2.5 \times \gamma \tag{2}$$

Let's take $y = SL/PH$,

$$y = 4.75 \pm \frac{\alpha}{2} + \frac{\beta}{50} - 2.5 \times \gamma \tag{3}$$

Note that this equation now expresses exactly a linear regression, showing pre-trained coefficient values (4.75, 0.5, 0.02, −2.5) modifying the independent variables α, β, γ. The coefficients shown are the ones obtained by training a linear regression model on our dataset. The predicted y can be multiplied by phone height to obtain a shirt length prediction.

4   Results

The comparison of performance given by Support Vector Regression (SVR), k-NN, Decision tree regression, and Linear regression is mentioned in Table 1. We use mean percentage error to report our results, since errors are to be calculated with reference to base measurements of various dimensions. Specifically, given observations y_i^j and predictions $\hat{y}_i^j$ for the ith measurement for the jth subject, we calculate the body part-specific error as (Fig. 5),

$$Error_i = \frac{1}{N} \sum_j \frac{|\hat{y}_i^j - y_i^j|}{y_i^j} \tag{4}$$

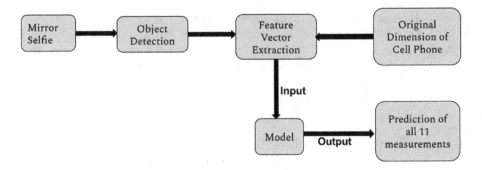

Fig. 5. Outline of testing phase of prediction model

Table 1. Comparison of % error between different regression models

Measurement	Linear regression	K-nn	SVR	Decision tree
Chest	4.88	4.58	6.09	9.04
Shoulder	3.24	3.81	4.16	5.71
Arm length	3.71	3.98	4.10	6.34
Wrist circumference	2.81	2.72	2.44	3.55
Collar	5.04	5.63	5.23	6.75
Pant length	2.34	2.51	2.98	3.64
Waist	6.73	5.84	6.09	9.28
Hip circumference	3.38	4.72	3.93	7.50
Thigh	4.24	4.02	4.38	5.96
Bottom of pant	4.42	3.93	3.08	5.49
Shirt length	3.82	4.42	3.64	3.70

Table 1 clearly shows the performance difference between all techniques for each measurement. Clearly, decision tree regression is not a good choice for predicting measurements. But the result shows that linear regression, SVR and k-NN are comparable. There is no single technique that outperforms the other two in all measurements. Linear Regression works better in case of predicting shoulder width, arm length, collar, pant length, and hip circumference. K-nearest neighbor works better in case of predicting chest, waist, and thigh, while SVR outperforms the other two in case of predicting wrist circumference, bottom of pant and shirt length prediction. But if we see the range of error in all measurements for these three models then k-NN seems a better choice. Linear Regression has the lowest error for pant length (2.34%) and the highest error for waist 6.73%. It has a range of 4.39%. Similarly, k-NN shows a range of 3.33 and for SVR it is 3.65%. So we can say that k-NN gives very stable results in compare to SVR and Linear Regression.

Keeping model interpretability in mind, using linear regression for all predictions appears to be the simplest and most stable solution across all 11 measurements. For this simple model, the average error is consistently lower than 5% across all 11 measurements.

5 Conclusion and Future Work

5.1 Conclusion

Body size prediction methods that involve the use of 3D scanning are highly accurate but, because of the logistics involved, are not feasible for e-commerce applications. As we review above, existing image-based body size prediction methods also require considerable logistical or computational efforts to set up correctly and obtain accurate results. We present a novel, extremely simple, system for obtaining reasonably accurate body size predictions using naturalistically obtained images, with no restrictions on pose, location and illumination conditions other than those endogenously applied by the physics of taking mirror selfies.

5.2 Limitations and Future Work

Our present empirical evaluation used only 70 male subjects, since female subjects felt reservations in sharing selfie photos with male researchers. This is a clear limitation of our present work, and should be addressed by future work that tests our model on a more diverse population. Whether this approach will work for females as is, or would require some changes in the feature representation is an important question to be addressed in future work.

We introduced a hand-crafted three variable feature vector and made predictions assuming that body part measurements strongly follow linear allometric scaling. Due to this strong assumption, our predictions for circumferential statistics are clearly biased, since we use a linear proportionality assumption where a quadratic one would be more mathematically appropriate. Obtaining more data with this human-computer interaction modality will eventually permit the construction of more complex models, and afford greater accuracy in prediction. Finally, in the definition of alpha and gamma we take the width of cell phone and width of a person as a width of the bounding box which surrounds them, but more sophisticated image processing may be tried to get exact width.

References

1. Zhu, S., Mok, P., Kwok, Y.: An efficient human model customization method based on orthogonal-view monocular photos. Comput. Aided Des. **45**(11), 1314–1332 (2013)
2. Meunier, P., Yin, S.: Performance of a 2D image-based anthropometric measurement and clothing sizing system. Appl. Ergon. **31**(5), 445–451 (2000)

3. Hung, P.C.-Y., Witana, C.P., Goonetilleke, R.S.: Anthropometric measurements from photographic images. Comput. Syst. **29**, 764–769 (2004)
4. Zhu, S., Mok, P.: Predicting realistic and precise human body models under clothing based on orthogonal-view photos. Procedia Manuf. **3**, 3812–3819 (2015)
5. Zhou, X., Chen, J., Chen, G., Zhao, Z., Zhao, Y.: Anthropometric body modeling based on orthogonal-view images. Int. J. Ind. Ergon. **53**, 27–36 (2016)
6. Tsoli, A., Loper, M., Black, M.J.: Model-based anthropometry: predicting measurements from 3D human scans in multiple poses. In: IEEE Winter Conference on Applications of Computer Vision, pp. 83–90 (2014)
7. Hirshberg, D.A., Loper, M., Rachlin, E., Black, M.J.: Coregistration: simultaneous alignment and modeling of articulated 3D shape. In: Fitzgibbon, A., Lazebnik, S., Perona, P., Sato, Y., Schmid, C. (eds.) ECCV 2012. LNCS, vol. 7577, pp. 242–255. Springer, Heidelberg (2012). https://doi.org/10.1007/978-3-642-33783-3_18
8. Liu, Z., Li, J., Chen, G., Lu, G.: Predicting detailed body sizes by feature parameters. Int. J. Cloth. Sci. Tech. **26**(2), 118–130 (2014)
9. Spahiu, T., Shehi, E., Piperi, E.: Extracting body dimensions from 3D body scanning. In: 6th International Conference of Textile, Tirana, Albania (2014)
10. Markiewicz, L., Witkowski, M., Sitnik, R., Mielicka, E.: 3D anthropometric algorithms for the estimation of measurements required for specialized garment design. Expert Syst. Appl. **85**, 366–385 (2017)
11. Xiaohui, T., Xiaoyu, P., Liwen, L., Qing, X.: Automatic human body feature extraction and personal size measurement. J. Vis. Lang. Comput. **47**, 9–18 (2018)

Retinal Vessel Classification Using the Non-local Retinex Method

A. Smitha⬛, P. Jidesh$^{(\boxtimes)}$⬛, and I. P. Febin

Department of Mathematical and Computational Sciences, National Institute
of Technology, Srinivasanagar, Mangalore 575025, Karnataka, India
jidesh@nitk.edu.in

Abstract. Automatic retinal vessel segmentation has turned out to be
highly propitious for medical practitioners to diagnose diseases like glaucoma and diabetic retinopathy. These diseases are classified based on
the thickness of the retinal vessel, the pressure imposed on the nerve
endings and optical disc to cup ratio of the retina. The state-of-the-art
device for this purpose presently available in the market is expensive and
has scope to meliorate sensitivity and precision of its performance. Thus,
automatic retinal blood vessel segmentation and classification is the need
of the hour. In this paper, a novel non-local total variational retinex based
retinal image preprocessing approach is proposed to extract the retinal
vessel features and classify the vessels using ground truth images. Matlab
implementation results indicate that an average accuracy of 94% with
an acceptable range of sensitivity and specificity could be achieved on
the retinal image database available online.

Keywords: Non-local total variational model · Vessel segmentation ·
Retinex framework · Supervised classification

1   Introduction

Retinal vessels engross bulk of the retinal area. The thickness and the strength
of the retinal vessels aid to identify diseases such as diabetic retinopathy, glaucoma, etc. Ophthalmoscopy or funduscopy is a preliminary examination done
by medical practitioners to contemplate the fundus of the eye. Initially, manual
segmentation of the retinal vessel was carried out by experts, which remains a
tedious, complicated task. With the technological advancements deep diving the
medical field in the present era, to expedite the medical examination process,
the Ophthalmoscope is equipped with automated technologies. IDx - DR is the
first FDA approved AI enabled autonomous optic device (in 2018) that detects
diabetic retinopathy. It has 87% Sensitivity and 90% Specificity [1]. This device
has drastically reduced the computational time complexity to less than a minute.
However, multiple factors impede accurate segmentation of these images such as
quality of lens in the ophthalmoscope, uneven illumination, low contrast, blurred
images, and so on. Thus, denoising the fundus images before feature extraction

© Springer Nature Switzerland AG 2020
U. S. Tiwary and S. Chaudhury (Eds.): IHCI 2019, LNCS 11886, pp. 163–174, 2020.
https://doi.org/10.1007/978-3-030-44689-5_15

is a daunting task and leaves room for further advancements in this field. A novel preprocessing method to extract retinal vessels using the non-local variational retinex algorithm is proposed in this paper.

The remaining content is organized as follows. Section 2 renders an overview of the existing literature. The proposed methodology is exemplified in Sect. 3. Section 4 manifests the result obtained, and the conclusion is presented in Sect. 5.

2   Literature Review

With the technological breakthrough spurred by machine learning, artificial intelligence, and deep learning, there are a plethora of research works related to fundus images. The traditional machine learning algorithms subsume unsupervised and supervised classification of retinal vessels [2]. SVM, Ensemble Classifiers, K-Means, and Naive Bayes methods are few among them. In these methods, generally, the images are preprocessed using enhancing and denoising algorithms. Preprocessing is followed by segmentation, feature extraction, and classification. Each of these methods has its pros and cons, and it can be tailored according to the application requirement and available resources. Contrary to the deep learning algorithms, these methods are efficacious when the available training data set and other resources are minimal. Considering the advantages of machine learning techniques, some of the related works and contributions are discussed here. Aslani et al. [3] presented a classification of retinal vessels using hybrid features. Memari et al. [4] proposed a supervised classification of retinal images to identify the vessel and non-vessel pixels using a matched filter and AdaBoost classifier. B- COSFIRE [5] and Frangi [6] are used for segmentation in their method. Even though promising results are showcased by their works, the preprocessing method has still scope for improvement. Nishaa et al. [7] discussed a novel method for identifying plus disease in retinopathy using adaptive segmentation. Thick and thin vessels are distilled separately and combined finally within the region of interest. Comparative analysis of results obtained is performed using SVM classifier. However, the sensitivity of the entire system could be enhanced by using other classification techniques. Wang et al. [8] illustrated a cascade based classification and retinal vessel extraction method. Here, multiple features are extracted using a matched filter, gray level based filter, DoG, and then feature dimensionality is reduced using PCA. Viraktamath et al. [9] demonstrated that morphological operations could be used to extract the blood vessels from fundus images. The performance of this model is assessed using multiple parameters. However, eliminating illumination inhomogeneity using non-local approach can augment the performance of these models.

Dey [10] discussed the effects of illumination and the need for accurate image correction and enhancement technique. Orlando [11] evaluated various preprocessing and feature extraction techniques. The author emphasized that the retinal images are subjected to illumination flaws and suggested intensity-based features, ridge detectors, line detectors, and wavelet operations as remedial measures to extract features from fundus images. Jidesh et al. [12] proposed a non-local variational framework for image denoising and image enhancement reckoning that an image is not only subjected to Gaussian noise but is also susceptible to data-dependent noise. Considering the advantages of the variational framework, the main objective of this paper is to extirpate the illumination inhomogeneity and intensify the retinal vessel pixels, extract the appropriate features of retinal vessels and classify the pixels using supervised machine learning algorithms.

3 Proposed Methodology

The overall proposed methodology for vessel extraction is depicted in Fig. 1 (a&b) and explicated below. Retinal images are acquired from the open-access database - DRIVE [13] and STARE [14]. DRIVE has 40 RGB fundus images, out of which 20 are catered for training with the ground truths (manual segmentation images) and 20 images are offered for testing purpose. STARE provides 20 images with two sets of manual segmentation, which can be utilized aptly for testing or training.

(a) The Overall methodology of retinal vessel classification

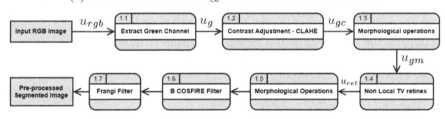

(b) Detailed steps of image preprocessing

Fig. 1. Steps involved in proposed retinal vessel classification

Let u_{rgb} represent appropriately resized, RGB image available in training data set of DRIVE/STARE as shown in Fig. 3(a). The green channel (u_g) is extracted (Fig. 2(c)) and is used for further processing as it yields the highest information [4]. The contrast of the image is adjusted using CLAHE - Contrast Limited Adaptive Histogram Equalization. The output u_{gc} of this stage is depicted in Fig. 3(d). Subsequently, the morphological operations like opening, top hat, and bottom hat are performed on the image, which is depicted in

Fig. 3(e) and represented as u_{gm}. Morphological processing is a set of non-linear operations performed by applying a structural element to the image based on the shape. These rectify the low contrast issues and reinforce the blood vessel region within the FOV (Field of View) determined by the mask. Since STARE data set is deprived of mask images, it is constructed by thresholding with the average values of R, G, and B values. The above described steps are symbolically represented in Eq. (1).

$$\begin{aligned} u_{rgb} &= u_r + u_g + u_b, \\ u_{gc} &= CLAHE(u_g), \\ u_{gm} &= morph(u_{gc}). \end{aligned} \tag{1}$$

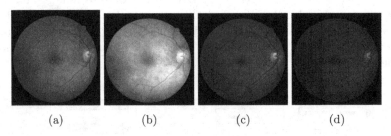

| (a) | (b) | (c) | (d) |

Fig. 2. Image and its components : (a) Composite RGB Image, (b) Red component, (c) Green Component and (d) Blue Component respectively.

3.1 Non-Local Total Variational Retinex

Retinex is a blend of two words - 'retina' and 'cortex'. The Retinex theory elucidates how our vision is consistent with the colour irrespective of the illumination factor. The goal of the retinex algorithm is to wipe out illumination inhomogeneity. According to the retinex theorem, any image 'U' defined on Ω can be expressed as the product of luminance (L) and reflectance (R) where $0 < R < 1$ and $0 < L < \infty$. In the log domain their product gets transformed to sum as shown in Eq. (2).

$$\begin{aligned} U &= L * R, \\ \log U &= \log L + \log R. \end{aligned} \tag{2}$$

Let $u_u = \log U, l_u = \log L$ and $r_u = \log R$. So, Eq. (2) becomes Eq. (3) :

$$u_u = l_u + r_u. \tag{3}$$

Contemplating smooth illumination function and non-smooth constant reflection function, the energy functional for the retinex model proposed in [15] is given as:

$$E(r_u, l_u) = \int_\Omega |\nabla r_u| + \frac{\gamma_1}{2} \int_\Omega |\nabla l_u|^2 dx + \frac{\gamma_2}{2} \int_\Omega (l_u - r_u - u_u)^2 dx + \frac{\gamma_3}{2} \int_\Omega l_u^2 dx, \tag{4}$$

where the first term on the right-hand side of Eq. (4) corresponds to TV norm of reflectivity, the second term represents L_2 norm of illuminance, the third term typifies the fidelity, and the last term instantiates a theoretical setting. Furthermore, $\gamma_1, \gamma_2, \gamma_3$ are positive scalars which denote the regularization parameters. Contrary to the earlier spatially smooth assumption of reflectance, here authors assume the reflectance to be piece-wise constant and therefore, use a TV norm instead of L^2 norm.

The use of TV norm has proved to give better results, but it serves as a mediocre approach when it comes to the preservation of image details. Therefore, the model shown above is reformulated using a non-local retinex framework under a variational formulation. The above model preserves detailed structures while subsidizing the inhomogeneity artefacts in the data. Several variants for the variational retinex method have been proposed such as [16] for extrapolating retinex using a non-local framework to retain more details. An improved model is being proposed in this work to enhance the visual quality of the output by employing a weberized TV [17] under non-local retinex framework as the regularizer along with a L^2 norm of the luminance as a constraint and data fidelity. Here, γ_1, γ_2 are scalar control parameters. The functional in Eq. (5) is optimized using a gradient descent algorithm to get the desired output.

$$E(r_u, l_u) = |\nabla r_u^w| + \gamma_1 ||\nabla l||_2^2 + \gamma_2 |u_u - l_u - r_u|. \tag{5}$$

In a Weberized TV method [17], a Weberized total variation supersedes the uniform total variation leveraging the influence of background intensity following Weber's law.[1] In the proposed method, the entire image 'U' is fragmented into multiple windows each of size, say, N × N. The similarity between patches within each window is assessed to obtain a weight matrix. The non-local weight (w_{nl}) is computed as the negative exponential of Euclidean distance as given in Eq. (6), which is discussed in [16]. Here, $u_u(x)$ and $u_u(y)$ correspond to the patch around pixel x and y, and G_σ represents a Gaussian kernel with standard deviation as σ.

$$w_{nl} = e^{\frac{-d(u_u(x), u_u(y))^2}{h^2}}, \tag{6}$$

where $d(u_u(x), u_u(y))^2 = \int_\Omega G_\sigma(t)(u_u(y + t) - u_u(x + t))^2 dt$. This approach penalizes less on the details while restoring the data. The parameters in each equation are amended according to the requirement to wangle a better quality of retinal vessels in the grayscale image. Under the above formulation, the model takes the form as Eq. (7), where u_{ret} represents retinex output and u_{gm} denotes the output of morphological operations.

$$u_{ret} = NLTVR(u_{gm}). \tag{7}$$

---

[1] Local gradient $|\nabla u_u| = \sqrt{u_{ux}^2 + u_{uy}^2}$ is transformed into weberized TV given as $|\nabla u_u|_w = \frac{|\Delta u_u|}{u_u}$ [17].

3.2 Segmentation and Feature Extraction

The output of NLTVR as shown in Fig. 3(f) is morphologically opened. The segmentation is performed by modifying the parameters of BCOSFIRE (Output is depicted in Fig. 3(g). Bar-selective combination of shifted filter responses (BCOSFIRE) is a combination of an asymmetric and a symmetric filters. The filter response is attained by convolving an image 'U(x, y)' with the difference of Gaussian (DoG(x, y)). DoG is a band-pass filter-based feature enhancement technique where the original image is subtracted from its blurred version. If the blurring function is a Gaussian kernel, the high-frequency spatial components of the image are abolished. In this method, the symmetric filter is applied first followed by the asymmetric filter.

The vessel segments acquired are further fortified using Frangi filter and the output of this is shown in Fig. 3(h). Frangi filter is based on the eigenvalues of the Hessian matrix which is computed using the second derivative of the input image. To ensure that both thick and thin vessels are extracted, the parameters of the Frangi filter for the DRIVE images are set as, $\beta_1 = 0.3$, $\beta_2 = 13$ and $\sigma = 1,1.1,1.2,1.3,.\ .\ .,2$. Since the segmentation method is adapted from [4], readers are encouraged to refer the same for an elaborate explanation of the segmentation method using BCOSFIRE and Frangi filter.

For classification, several patches are selected around a randomly selected pixel, and then 37 features are extracted as listed in Table 1. Minimum Redundancy Maximum Relevance (MRMR) algorithm is applied to select 10 best discriminating features. AdaBoost classifier predicts the vessel and non-vessel pixels using the reduced feature values. The AdaBoost classifier is an ensemble classification algorithm that is predominantly used for classification of the retinal images. To compare the results with the available literature, decision tree method and KNN is also implemented and the performance of the classifier is evaluated.

4 Results and Discussion

The proposed methodology is implemented using Matlab 2017b on core i5-6200U CPU @2.3 GHz processor with 4 GB RAM. Images obtained at the end of each processing step for a randomly selected image of the DRIVE database are depicted in Fig. 3(a–h). The preprocessed image is converted to binary using thresholding (see Fig. 3(i)) for better visual comparison with the ground truth image. A sample image from STARE image data set, it's corresponding ground truth image, computed mask, and vessels extracted using the proposed method are rendered in Fig. 5. The execution time for the segmentation of one image is 4.17 min. Sample values of the ten features selected using MRMR for the vessel and the non-vessel pixel patches are shown in Table 2. As evident from the values, the histogram and gray level run length based feature values remarkably distinguish retinal vessel pixels from background. To compare the average accuracy with all the works discussed in the literature review, approximately 15000 pixels are selected randomly from each image, and then the overall performance

Table 1. List of features extracted.

1. **Histogram based features**	(h) Inverse difference normalized	(v) Energy
(a) Standard Deviation	(i) Difference variance	3. **Gray level run length based features**
(b) Energy	(j) Dissimilarity	(a) Gray Level Non-Uniformity
(c) Skewness	(k) Difference entropy	
(d) Entropy	(l) Maximum probability	(b) Long Run Emphasis
(e) Kurtosis	(m) Information measure of correlation I	(c) Low Gray Level Run Emphasis
(f) Mean		(d) Short Run Emphasis
2. **GLCM based features**	(n) Sum average	
(a) Correlation 1	(o) Sum variance	(e) Run Percentage
(b) Autocorrelation	(p) Information measure of correlation II	(f) High Gray Level Run Emphasis
(c) Contrast		(g) Run Length Non-Uniformity
(d) Cluster Shade	(q) Sum of squares	
(e) Cluster Prominence	(r) Sum entropy	4. **Gabor features**
(f) Homogeneity 1	(s) Correlation 2	(a) squared energy
(g) Inverse difference moment normalized	(t) Entropy	(b) amplitude
	(u) Homogeneity 2	

is measured using Decision Tree Classifier, AdaBoost classifier, and KNN classifier. As evident from the graph shown in Fig. 4, Decision Tree classifier touches the accuracy greater than 95% for the DRIVE images while AdaBoost classifier serves better for STARE images.

The classifier performance is measured using Accuracy, Sensitivity, Specificity, and Dice Coefficient that are estimated using the following formulae.

1. Accuracy $= \frac{TTP+TTN}{TTP+TFP+TTN+TFN}$.
2. Sensitivity $= \frac{TTP}{TTP+TFN}$.
3. Specificity $= \frac{TTN}{TTN+TFP}$.
4. Dice Coefficient $= \frac{2*TTP}{2*TTP+TFN+TFP}$.

where TTP, TTN, TFP, TFN represents Total True Positive, Total True Negative, Total False Positive and Total False Negative respectively, and are obtained from the confusion matrix. Sensitivity and specificity are the measure of true positive and true negative rate respectively. In retinal vessel extraction, true positive refers to correct classification of the vessel pixels and true negative denotes the correct classification of non vessel pixels. Apparently, the accuracy approximates to the average of sensitivity and specificity. Table 3 summarizes the results obtained from the DRIVE and STARE data set. It projects that the segmentation of retinal vessels acquired using non-local variational framework is in the acceptable range compared to the existing works. Cross performance of the classifier is evaluated by training the classifier using DRIVE images and

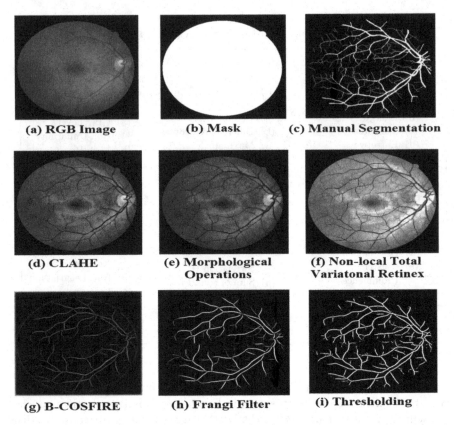

(a) RGB Image   (b) Mask   (c) Manual Segmentation

(d) CLAHE   (e) Morphological   (f) Non-local Total
                       Operations   Variatonal Retinex

(g) B-COSFIRE   (h) Frangi Filter   (i) Thresholding

Fig. 3. Images obtained during pre-processing of one image chosen from the DRIVE
database

testing it on the STARE images and vice-versa. The Receiver Operating Characteristic (ROC) curve obtained is projected in Fig. 6. ROC is a plot of the true positive rate against the false positive rate. In a binary classification, always a higher true positive rate and a lower false positive rate is desirable. Therefore, the closer an ROC curve is to the upper left corner, the more efficient is the binary classification of vessel and non-vessel pixels. The graph in Fig. 6 depicts that the classifier performs better when it is trained using DRIVE and tested on STARE images. The performance of the proposed model for segmentation is measured using the Area Under a ROC Curve (AUC) which must be close to 1, to indicate all pixels are precisely segmented following the ground truth images. The AUC values obtained using the proposed method is 0.97 for both the DRIVE and STARE databases. Table 4 presents AUC, dice coefficient, accuracy, sensitivity, and specificity values obtained for various patch size using AdaBoost classifier for the proposed method and the method described in [4]. Five fold cross-validation is incorporated to optimize the classifier by considering 90% of all the images from DRIVE and STARE dataset for training and 10% of the

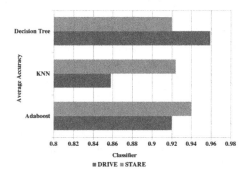

Fig. 4. Comparison of accuracy obtained using various classifiers

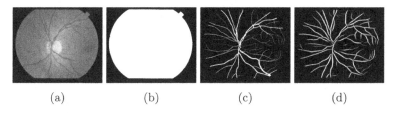

(a)   (b)   (c)   (d)

Fig. 5. Images obtained after pre-processing one image chosen from STARE database: (a) RGB Image, (b) Computed Mask, (c) Manual Segmentation, (d) Segmentation by the proposed method

Table 2. Sample feature values extracted to distinguish vessel pixel from a non-vessel pixel considering a patch size of 15×15 around a pixel

Feature	Non-vessel patch	Vessel patch
High gray-level run emphasis	62	48.3977
Variance	0	3.2579
Run length non-uniformity	0.36686	1.59171
Entropy	1	0.5442
Cluster prominence	0	542.2619
Run percentage	24.8387	98.16728
Sum average	2	4.39743
Long run emphasis	127.064	13.75092
Low gray-level run emphasis	1	19.9665
Information measure of correlation1	0	0.46414

images for testing. From Table 4, the proposed model outperforms the existing works for patch size beyond 11 × 11. The accuracy recorded using KNN classification algorithm is depicted in the graph (Fig. 7) for various 'k' values ranging from 5 to 50. Seemingly, the graph reveals that patch sizes beyond 9 × 9 exhibits approximately equal accuracy levels for varying size of 'k'. However, the average accuracy level achieved is less than the accuracy achieved using AdaBoost Classifier.

Table 3. Comparison of average accuracy, sensitivity, specificity and AUC for the proposed method with existing literature works

Method for DRIVE images	Accuracy	Sensitivity	Specificity	AUC
Proposed Method	**0.9589**	**0.9667**	0.9511	0.97
Hashemzadeha et al. [2]	0.9531	0.7830	0.9800	**0.9752**
Aslani et al. [3]	0.9513	0.7545	0.9801	0.9682
Memari et al. [4]	0.8399	0.8726	0.9884	0.9795
Nishaa et al. [7]	0.9383	0.73851	**0.98604**	0.9614
Xiaohong et al. [8]	0.9541	0.7648	0.9817	-
Viraktamath et al. [9]	0.9586	0.6302	0.9859	-
Method for STARE images	Accuracy	Sensitivity	Specificity	AUC
Proposed method	0.9428	**0.9252**	0.9599	0.97
Hashemzadeha et al [2]	**0.9691**	0.8087	**0.9892**	**0.9853**
Aslani et al. [3]	0.9605	0.7556	0.9837	0.9789
Memari et al. [4]	0.9514	0.8085	0.9798	0.9701
Xiaohong et al. [8]	0.9640	0.7523	0.9885	-

Table 4. Accuracy of the classifier for randomly selected patches of different size from fundus images.

Patch size	Proposed method					As mentioned in [4]		
	AUC	Dice coefficient	Accuracy	Sensitivity	Specificity	Accuracy	Sensitivity	Specificity
3 × 3	0.8	0.766234	0.7916	0.746835	0.774647	0.932100	0.812425	0.950528
5 × 5	0.85	0.826689	0.84479	0.82739	0.82629	0.922397	0.804419	0.939632
7 × 7	0.93	0.887469	0.89792	0.949127	0.851456	0.900553	0.796018	0.915613
9 × 9	0.91	0.798177	0.83126	0.880885	0.769275	0.884507	0.814860	0.894425
11 × 11	0.93	0.85533	0.87632	0.93759	0.81601	0.864167	0.812207	0.871491
13 × 13	0.95	0.836727	0.86258	0.901462	0.806584	0.847372	0.807586	0.852919
15 × 15	0.95	0.84369	0.88437	0.888315	0.820505	0.829701	0.807140	0.832814
17 × 17	0.96	0.820314	0.884375	0.830543	0.814839	0.810445	0.810933	0.810378
19 × 19	0.96	0.839283	0.881134	0.84395	0.836466	0.792581	0.805950	0.790771
21 × 21	0.97	0.85054	0.905208	0.848425	0.851926	0.778764	0.797988	0.776182
23 × 23	0.97	0.863719	0.916667	0.858177	0.867612	0.764832	0.793468	0.761018
25 × 25	0.97	0.867763	0.90729	0.856494	0.876027	0.752695	0.779191	0.749188

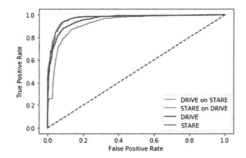

Fig. 6. ROC curve plot

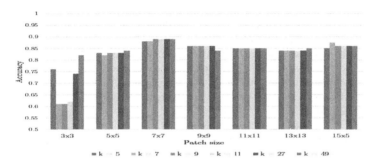

Fig. 7. Accuracy measure using KNN classifier

5   Conclusion

In this paper, a novel non-local variational retinex method is proposed to extract the retinal vessels more efficiently by reducing the inhomogeneity and noise intervention in the input images using a supervised classification method. The experimental analysis has shown that the proposed model performs consistently well to maintain higher sensitivity and specificity. The extracted information about the retinal vessels serves as a baseline, and the succeeding works include assessing the diseases like diabetic retinopathy and glaucoma using machine learning algorithms.

Acknowledgements. Dr. Jidesh and Ms. Febin would like to thank the Science and Engineering Research Board, India for providing financial support under the Project Grant No. ECR/2017/000230. Ms. Smitha expresses her gratitude to the Ministry of Human Resource, Government of India for providing financial support for pursuing Ph.D. at National Institute of Technology Karnataka, Surathkal. Furthermore, the authors acknowledge the authors of [4–6] for making the dataset available for testing and providing the Matlab code for comparison purpose.

References

1. Autonomous AI that instantly detects disease. https://www.eyediagnosis.co/idx-dr-eu-1
2. Hashemzadeha, M., Azarb, B.A.: Retinal blood vessel extraction employing effective image features and combination of supervised and unsupervised machine learning methods. Artif. Intell. Med. **95**, 1–15 (2019)
3. Aslani, S., Sarnel, H.: A new supervised retinal vessel segmentation method based on robust hybrid features. Biomed. Signal Process. Control **30**, 1–12 (2016)
4. Memari, N., Ramli, A.R., Bin Saripan, M.I., et al.: Supervised retinal vessel segmentation from color fundus images based on matched filtering and AdaBoost classifier. PLoS ONE **12**(12), e0188939 (2017). https://doi.org/10.1371/journal.pone.0188939
5. Azzopardi, G., Strisciuglio, N., Vento, M., Petkov, N.: Trainable COSFIRE filters for vessel deldineation with application to retinal images. Med. Image Anal. **19**(1), 46–57 (2015). https://doi.org/10.1016/j.media.2014.08.002. PMID: 25240643
6. Frangi, A.F., Niessen, W.J., Vincken, K.L., Viergever, M.A.: Multiscale vessel enhancement filtering. In: Wells, W.M., Colchester, A., Delp, S. (eds.) MICCAI 1998. LNCS, vol. 1496, pp. 130–137. Springer, Heidelberg (1998). https://doi.org/10.1007/BFb0056195
7. Nishaa, K.L., Sreelekhaa, G., Sathidevi, P.S., et al. : A computer-aided diagnosis system for plus disease in retinopathy of prematurity with structure adaptive segmentation and vessel based features. Comput. Med. Imaging Graph. J. 72–94 (2019). Springer, Heidelberg (2016). https://doi.org/10.10007/1234567890
8. Xiaohong, W., Jiang, X., Ren, J.: Blood vessel segmentation from fundus image by a cascade classification framework. Pattern Recogn. **88**. https://doi.org/10.1016/j.patcog.2018.11.030
9. Viraktamath, S.V., Koti, V., Ragi, S., et al.: Blood vessels extraction of retinal image using morphological operations. In: Proceedings on: The International Conference on Inventive Research in Computing Applications (ICIRCA 2018) (2018). IEEE Xplore Compliant Part Number: CFP18N67-ART; ISBN: 978-1-5386-2456-2
10. Dey, N.: Uneven illumination correction of digital images: a survey of the state-of-the-art. Optik - Int. J. Light Electron Opt. **183**, 483–495 (2019)
11. Orlando, J.I., del Fresno, M.: Reviewing Preprocessing and Feature Extraction Techniques for Retinal Blood Vessel Segmentation in Fundus Images. Mecánica Computacional (2014)
12. Jidesh, P., Shivarama Holla, K.: Non-local total variation regularization models for image restoration. Comput. Electr. Eng. **67**, 114–133 (2018)
13. DRIVE: Digital Retinal Images for Vessel Extraction available. https://www.isi.uu.nl/Research/Databases/DRIVE/
14. STARE: STructured Analysis of the Retina. http://cecas.clemson.edu/~ahoover/stare/
15. Ng, M.K., Wang, W.: A total variation model for retinex. SIAM J. Imaging Sci. **4**(1), 345–365 (2011)
16. Zossoy, D., Tran, G., Osher, S.J.: Non - local retinex - a unifying framework and beyond. SIAM J. Imaging Sci. **8**(2), 787–826 (2014)
17. Shen, J.H.: On the foundations of vision modeling: Weber's law and Weberized TV restoration. Physica D **175**(3–4), 241–251 (2003)

Rule Generation of Cataract Patient Data Using Random Forest Algorithm

Mamta Santosh Nair[1](✉) and Umesh Kumar Pandey[2](✉)

[1] MATS University, Raipur, India
mamtanair@yahoo.com
[2] MSIT, MATS University, Raipur, India
Umesh6326@gmail.com

Abstract. Cataract is one of the common problems among the humans. Cataract is the condition caused due to clouding of lens in the eye which eventually may lead to blindness. In last few years, data mining has been widely used to build the predictive model in various fields. In this paper, historical data of cataract patient has been used to build the predictive model. Random forest algorithm is one of the decision tree algorithms for predictive modeling. Random forest algorithm incorporates advantages of classification and regression. Present study uses random forest method to create a model for prediction of cataract. The random forest algorithm is also tested for Out of Bag estimation error.

Keywords: Data mining · Classification · Random forest · Rule generation · Out of bag error · Decision support system

1 Introduction

Decision making is a tough job. One decision relies on many factors. In data mining algorithms, decision tree algorithms are one of the most widely used algorithms for predictions. Random forest algorithm is one of such robust and simplistic algorithm that works on ensemble learning method.

Data mining is applied in medical science, astronomy and other field to extract information from the data set. This data set has large number of attribute and complexity of inferring information.

One of the major causes of blindness in the world is cataract. Cataract is the preventable blindness if the patient is operated in time. Several organizations worldwide are working towards spreading the word about cataract and also about surgeries performed for cataract. According to World Health Organization, Cataract is responsible for 51% of world blindness [15]. World Health Organization defines this condition as "Cataract is clouding of the lens of the eye which impedes the passage of light. Although most cases of cataract are related to the aging process, occasionally children can be born with the condition, or a cataract may develop after eye injuries, inflammation, and some other eye diseases" [15]. The statistics collected from many agencies and previous literatures are serious enough to take a giant step towards preserving the vison. Data of cataract

© Springer Nature Switzerland AG 2020
U. S. Tiwary and S. Chaudhury (Eds.): IHCI 2019, LNCS 11886, pp. 175–188, 2020.
https://doi.org/10.1007/978-3-030-44689-5_16

patients' needs to be studied and analyzed so as to reveal the hidden trends which can further be used to create awareness among general population.

In this paper, we have collected the data of patients with eye problems among which many are suffering from cataract. And the collection also includes other details of patients like dietary habits, addictions, living environment etc. which may help to predict the chances of getting cataract. Data mining algorithms carry out this assessment to assist in the decision making process.

2 Literature Review

Data mining can help see us what is not directly visible but is underlying the obvious. It finds out the pearls of patterns and trends from the oceans of data. Data mining performs analysis of information to find possible outputs [3]. The methods where the hidden trends of data are identified, analyzed and then categorized into useful knowledge is known as Data Mining [4]. It finds patterns or trends, which are interesting and useful too. It helps to see beyond all the knowledge. And finally, it allows one to decide upon facts and predict the classes. Data Mining can play a significant role in arranging the data into different classes [6].

Decision tree algorithm breaks the dataset multiple times from top to bottom app-roach and then later horizontally at the same level till all the data items belonging to a class are identified [5]. A decision tree structure is made of root, internal and leaf nodes. Most decision tree classifiers perform grouping or classification in two steps: firstly, a tree is grown fully and then shortening or trimming of trees are done. The tree is grown from the top first then it is divided further into branches till all class labels are identified. While trimming process is carried on, a tree is cut wherever required to improve the accuracy. The trimming begins from lowermost node [10].

A decision tree is like a flowchart in structure and layout where every inner node represents a condition on an attribute and each branch represents a yes/no result of the condition and class label is represented by each leaf node (or terminal node). The leaf node is the last node. Classification rules are generated going from the top node to the terminal node of the decision tree [2].

Classification algorithm learns in supervised environment. It finds out and allocates class labels to data items by applying the already acquired knowledge of class which the data records belong [1]. Classification technique can be solving several problems in different fields like medicine, industry, business, and science. Basically it involves finding rules that categorize the data into disjoint groups [14].

The objective of the classification is to build a model based on some example cases with some attributes to describe the objects or one attribute to describe the group of the objects. Then, the model is used to predict the group attributes of new cases from the domain based on the values of other attributes [12].

Classification is the step wise process of finding a set of models which describe and performs allocation of data classes. The derived model is based on the analysis of a set of training data (i.e. data objects whose class label is known) [13].

Random Forest algorithm is a classifier model consisting of collection of trees or jungle like appearance where independent random vectors are distributed identically and every tree ends terminally for the accurate class [8]. At each step new random vector is generated which is independent of the previous random vectors with same distribution and then forms a tree using the training set [9]. Random Forest uses decision Trees as base classifier. This ensemble learning method is used for classification and regression of data. An ensemble consists of number of trained models whose predictors are combined to classify new variables.

Random forests are an effective tool in prediction. Because of the Law of Large Numbers, they do not overfit. It inserts just the right amount of randomness and we get good and accurate classifiers and regressors [7]. The random selection of dimensions to choose the splitting variable can be done as well as the choice of coefficients for random combinations of features [11].

Nayer [18] did his research work on diabetes mellitus detection using machine learning. Stacking ensemble method used in this research work built upon linear discriminant analysis, recursive tree and KNN.

Beaulac and Rosenthal [19] studied undergraduate students of Canada university in past 10 years using random forest. Using random forest, they identified most important variable useful to the classifier that reveals information for the university administration.

Sugandhi, Yasodha, Kannan [20] used five classification algorithms for prediction of cataract. The algorithms used by them were Naïve Bayes, SMO, J48, REP Tree and Random Tree. Authors also found mean absolute error and correctly classified instance generated by all the algorithms. They found random forest algorithm to be most accurate classifier with prediction accuracy at 84.87%.

Niya [21] developed automatic cataract detection methodology. The methodology involved pre-processing, feature extraction and classification. SVM classifier was used for prediction of cataract and regression method used for grading of cataract.

3 Data Collection and Research Instrument

The research work uses cataract patient data for the study. Dataset used in this research work is primary data collected through questionnaire. Questionnaire has been designed in consultation with Ophthalmologists. The questionnaire has also been designed considering the factors responsible for cataract as per specified by World Health Organization website. World Health organization mentions smoking, diabetes mellitus, exposure to ultra violet rays and high body mass index to be some of the cataract causing parameters [15]. Keeping in view of all factors total 43 different parameters selected for the data collection. These parameters included personal details, food habits, medical and birth history and addictions etc. The target location of the data collection is Raigad District of Maharashtra, India. Questionnaire was prepared in English and Marathi language. This questionnaire distributed among the cataract patient of approximately 700. Because of low education, most of the respondents are not familiar with the questionnaire system, thus assistance provided for the form filling. The data include people of both genders of different age groups. The data also had good mix of rural including tribal as well as urban population. Total approx. 500 forms received and filled at the camps and outpatient department (OPD) of doctors. Only 297 forms found complete and were selected for analysis. Certain parameters in questionnaire have received no answers or very less amount of entries. Thus, those attributes were removed from the dataset and only 17 attributes were considered for the study.

From the dataset attribute 'cataract' is used as a class name and other 16 variables are predictor variable. The dataset is studied in R software for performing random forest algorithm. R has inbuilt packages for random forest. Packages used in this study are "randomForest", "dplyr", "readxl" and "reprtree". Table 1 represents Attribute name and symbolic name used in the code development and to increase the visibility of the tree.

Table 1. Attribute names and abbreviations

Symbolic name	Attribute name	Symbolic name	Attribute name	Symbolic name	Attribute name
A	Age	G	Addiction	M	Occupation history in years
B	Gender	H	Hypertension duration	N	Sun exposure in hours
C	Occupation	I	Diabetes duration	O	History of trauma (yes/no)
D	Height	J	Cholesterol duration	P	Spectacle use duration
E	Weight	K	Surgical history (yes/no)	Q	Cataract (yes/no)
F	Diet	L	Type of surgery		

4   Importance of Attributes

One of the most robust characteristics provided by Random Forest is the importance factor of attributes. Table 2 gives the list of attributes with their importance in Class 1 and Class 2. Table 2 also gives the Mean Decrease Accuracy and Mean Decrease Gini. Mean Decrease Accuracy is where values of variables are randomly permuted and it is also known as permutation importance.

Mean decrease Gini is also known as Gini importance. The mean decrease in Gini coefficient is a measure of how each variable contributes to the homogeneity of the nodes and leaves in the resulting random forest. Each time a particular variable is used to split a node, the Gini coefficient for the child nodes are calculated and compared to that of the original node. The Gini coefficient is a measure of homogeneity from 0 (homogeneous) to 1 (heterogeneous). Attributes with a large mean decrease in accuracy are more important for classification of the data. In the given Table 2 attribute age is highest important with the Meandecrease accuracy at 29.2196387 followed by attribute type of surgery at 17.944127 and so on. Similarly, Meandecreasegini is highest for attribute age at 37.377705 followed by attribute weight and so on.

Table 2. Importance of attributes

Symbolic name	1	2	Mean Decrease Accuracy	Mean Decrease Gini
a	32.11137067	10.9070787	29.2196387	37.377705
b	0.04448175	1.3997418	1.1273503	1.243247
c	2.54318416	6.4052249	6.9375734	12.35707
d	0.33578291	1.7408604	1.4615954	13.086229
e	−0.83469508	6.668167	4.5231379	20.628917
f	1.16039169	3.4639696	3.1889527	2.61541
g	4.55669683	3.0163488	4.8897378	2.234993
h	7.7225398	0.4988208	6.067578	8.993176
i	7.57758079	−1.6094264	4.6023365	5.432224
j	0.78475064	−0.3456008	0.1164765	1.35444
k	7.65858699	2.7526541	7.989449	2.167219
l	20.67840801	2.376934	17.944127	6.086469
m	7.02724257	−0.3946136	4.6610465	8.43427
n	0.10091254	8.399354	5.9438672	9.882421
o	−0.51346518	6.1557188	3.4134775	2.085709
p	0.84577193	−0.3872764	0.2524296	13.915559

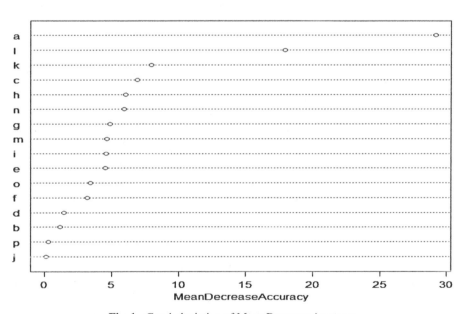

Fig. 1. Graph depiction of Mean Decrease Accuracy

The graph plotted for the variable importance is shown in Figs. 1 and 2. The plot shows each variable on the y-axis, and their importance on the x-axis. Attributes are ordered top-to-bottom as most- to least-important. Therefore, the most important attributes are at the top and an estimate of their importance is given by the position of the dot on the x-axis. Three least important variables were removed but OOB estimation error increased after removing it. Random forest algorithm used all 16 variables for rule generation.

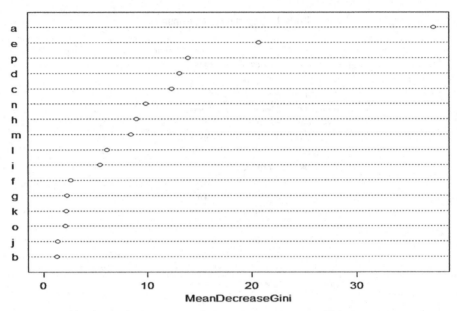

Fig. 2. Graph depiction of Mean Decrease Gini

5   OOB Estimation Error

The out-of-bag error is an error estimation technique often used to evaluate the accuracy of a random forest and to select appropriate values for tuning parameters, such as the number of candidate predictors that are randomly drawn for a split, referred to as mtry [16]. For each observation zi = (xi, yi), construct its random forest predictor by averaging only those trees corresponding to bootstrap samples in which zi did not appear [17]. The out of bag estimate chooses all the samples which were left during the tree creation and error is estimated for that sample.

6   Analysis and Discussions

Random forest generates an OOB error estimation depending upon the seed value and mtry. Table 3 shows OOB error estimation at different seed value and mtry. Random forest code run in R, for different values of mtry ranging from 3 to 13 and the value of set.seed was from 1 to 10. Set.seed is used for staring point of random number generation and mtry tuning parameters. Thus total 130, OOB estimation recorded as shown in Table 3.

Among the obtained OOB error estimation values, lowest value (30.98) is obtained at set.seed value 3 and mtry value 12, which is used for further study and generating rule. The R Code generates the 500 trees and selecting 12 variables at each spilt. Total 297 records considered for developing the model.

Table 3. OOB error estimation at set.seed and mtry

		mtry										
		3	4	5	6	7	8	9	10	11	12	13
Set.seed	1	31.31	33	32.66	34.01	34.34	34.01	34.34	32.32	33.33	34.01	33.67
	2	33.33	34.01	33.33	32.32	32.66	33.67	34.68	34.01	34.34	34.01	32.66
	3	33.67	35.02	32.32	35.02	33.67	32.32	35.35	34.01	34.68	**30.98**	34.01
	4	34.68	34.01	33	32.66	34.34	34.34	35.02	34.68	32.32	33.67	35.69
	5	31.65	32.32	35.02	33.67	34.68	34.01	31.99	33	33	31.99	33.67
	6	32.32	32.66	33.67	33.67	34.34	33	33.33	35.69	35.02	35.02	33.33
	7	33.33	34.01	31.65	32.66	35.35	33.33	35.35	36.03	35.35	35.35	34.68
	8	33	34.34	31.99	34.01	32.66	33.67	34.01	33.33	34.68	33.67	34.68
	9	31.99	33	33.33	34.01	35.35	33.33	34.34	34.68	35.69	34.34	35.02
	10	32.66	34.01	33.67	33.67	34.01	33	34.34	33	31.99	34.01	34.01

Table 4 shows confusion matrix. Confusion matrix shows that total 92 records are correctly classified into class 1 whereas 54 records are wrongly classified. Similarly, 113 records classify correctly into the class 2 whereas 38 records are wrongly classified. Classification error for the class 1 is 0.3698 and for the class 2 classification error is 0.2516556.

Table 4. Confusion matrix

N = 297	1	2	Total	Class. error
1	92	54	146	0.3698630
2	38	113	151	0.2516556
Total	130	167		

Accuracy related various parameters calculated from the confusion matrix are given in the Table 5 as diagnostic testing of accuracy. In Table 5 accuracy of classification of the tree reported 69.0326%. Another important point to identify the accuracy is precision and recall. Precision and recall both collectively represent detailed picture of accuracy. At one side precision represents relevancy whereas recall represent correctness of the model. Precision value of the decision tree is 67.6646% whereas recall is 74.8344%. Precision value explains 67.6646% of positive identification actually correct and recall explains that 74.8344% actual positives identified correctly. Misclassification rate of the decision tree is 28.9562%. Epidemiologist and other use prevalence which is in contrast incidence measure new cases in the population. Point prevalence reported in the Table 5 is 50.8417%. Prevalence explains that reported percentage of people are having condition of cataract at the time of collection of data. False positive rate i.e. 36.986% condition

Table 5. Parameters obtained from the values of confusion matrix

Parameter	Value	Parameter	Value	Parameter	Value
Accuracy of classifier	0.6902356	Precision	0.676646	Recall	0.748344
Misclassification rate	0.289562	Prevalence	0.508417	False positive rate	0.369863
F-score	0.710691	True negative rate	0.63013		

improperly exist. True negative rate is 63.013 reported in the Table 5 explains that actual nonexistence of condition is correctly classified. F score indicator represents harmonic mean between precision and recall. F score value is reported in Table 5 is 71.0691% represents similarity between the groups.

Table 6 shows the database of tree generation. Sr. no. shows the node number. Left daughter column indicates the node number which is associated with the left part of the splitting node. Right daughter indicates that which node number is associated with right part of the splitting node. Split var indicates the name of variable which is used for the splitting. Status column indicates that whether node is terminal or non-terminal node. If the status is 1 it means it is non-terminal and −1 status indicates that it is terminal node and indicates class name. Predication column shows the name of the class. <NA> in prediction column indicates that node is not leaf node and has further left or right or both sub-trees.

Table 6 enlists the rules generated by random forest. First column of the table represents node of the tree. Second column highlights left child of the current node. Third column represents right child of the current node. Fourth column represents code name of the splitting variable. Splitting code is defined in the Table 2. Fifth column is the split point that represents threshold value. Continuous values less than goes to the left side of the tree and greater than and equal into right side of the tree; in case of categorical variable respective values are mentioned in the column. Column 6 is status represents whether the current node belongs to leaf node. Status 1 represents non-leaf node whereas

Table 6. Rules generated by Random Forest

Node	Left daughter	Right daughter	Split var	Split point	Status	Prediction
1	2	3	A	54.5	1	<NA>
2	4	5	L	2	1	<NA>
3	6	7	G	1	1	<NA>
4	8	9	E	49	1	<NA>
5	10	11	C	167	1	<NA>
6	12	13	P	21	1	<NA>

(continued)

Table 6. (*continued*)

Node	Left daughter	Right daughter	Split var	Split point	Status	Prediction
7	14	15	H	6.5	1	\<NA\>
8	0	0	\<NA\>	0	−1	1
9	16	17	F	1	1	\<NA\>
10	18	19	H	0.5	1	\<NA\>
11	20	21	D	5.4	1	\<NA\>
12	22	23	C	239	1	\<NA\>
13	24	25	A	58.5	1	\<NA\>
14	26	27	A	58.5	1	\<NA\>
15	28	29	P	5	1	\<NA\>
16	30	31	P	8.5	1	\<NA\>
17	0	0	\<NA\>	0	−1	2
18	32	33	P	18	1	\<NA\>
19	34	35	P	2.5	1	\<NA\>
20	36	37	N	3	1	\<NA\>
21	0	0	\<NA\>	0	−1	1
22	38	39	B	1	1	\<NA\>
23	0	0	\<NA\>	0	−1	1
24	0	0	\<NA\>	0	−1	2
25	0	0	\<NA\>	0	−1	1
26	40	41	H	1	1	\<NA\>
27	0	0	\<NA\>	0	−1	2
28	0	0	\<NA\>	0	−1	1
29	0	0	\<NA\>	0	−1	2
30	0	0	\<NA\>	0	−1	1
31	0	0	\<NA\>	0	−1	2
32	42	43	K	1	1	\<NA\>
33	44	45	D	5.4	1	\<NA\>
34	0	0	\<NA\>	0	−1	2
35	46	47	E	74	1	\<NA\>
36	0	0	\<NA\>	0	−1	2
37	48	49	P	7.5	1	\<NA\>
38	50	51	I	1.5	1	\<NA\>

(*continued*)

Table 6. (*continued*)

Node	Left daughter	Right daughter	Split var	Split point	Status	Prediction
39	52	53	L	2	1	<NA>
40	54	55	N	1.5	1	<NA>
41	0	0	<NA>	0	−1	2
42	56	57	N	7	1	<NA>
43	58	59	P	9.5	1	<NA>
44	0	0	<NA>	0	−1	1
45	0	0	<NA>	0	−1	2
46	0	0	<NA>	0	−1	1
47	60	61	I	5	1	<NA>
48	0	0	<NA>	0	−1	1
49	0	0	<NA>	0	−1	2
50	62	63	D	5.25	1	<NA>
51	0	0	<NA>	0	−1	1
52	0	0	<NA>	0	−1	1
53	64	65	E	76	1	<NA>
54	0	0	<NA>	0	−1	1
55	0	0	<NA>	0	−1	2
56	0	0	<NA>	0	−1	1
57	66	67	E	76.5	1	<NA>
58	0	0	<NA>	0	−1	1
59	0	0	<NA>	0	−1	2
60	0	0	<NA>	0	−1	2
61	0	0	<NA>	0	−1	1
62	68	69	D	4.5	1	<NA>
63	70	71	A	56.5	1	<NA>
64	72	73	N	0.5	1	<NA>
65	74	75	A	67	1	<NA>
66	0	0	<NA>	0	−1	2
67	0	0	<NA>	0	−1	1
68	76	77	C	1	1	<NA>
69	78	79	K	1	1	<NA>
70	0	0	<NA>	0	−1	1

(*continued*)

Table 6. (*continued*)

Node	Left daughter	Right daughter	Split var	Split point	Status	Prediction
71	80	81	A	79	1	<NA>
72	0	0	<NA>	0	−1	1
73	0	0	<NA>	0	−1	2
74	0	0	<NA>	0	−1	2
75	0	0	<NA>	0	−1	1
76	0	0	<NA>	0	−1	1
77	0	0	<NA>	0	−1	2
78	82	83	O	1	1	<NA>
79	84	85	D	5.05	1	<NA>
80	86	87	N	0.5	1	<NA>
81	0	0	<NA>	0	−1	1
82	0	0	<NA>	0	−1	2
83	88	89	E	50	1	<NA>
84	0	0	<NA>	0	−1	2
85	90	91	P	0.5	1	<NA>
86	0	0	<NA>	0	−1	1
87	92	93	E	71	1	<NA>
88	0	0	<NA>	0	−1	2
89	0	0	<NA>	0	−1	1
90	0	0	<NA>	0	−1	2
91	94	95	H	5	1	<NA>
92	96	97	E	63.5	1	<NA>
93	0	0	<NA>	0	−1	2
94	0	0	<NA>	0	−1	1
95	0	0	<NA>	0	−1	2
96	98	99	E	58.5	1	<NA>
97	0	0	<NA>	0	−1	1
98	100	101	N	1.5	1	<NA>
99	0	0	<NA>	0	−1	2
100	102	103	E	53.5	1	<NA>
101	0	0	<NA>	0	−1	1
102	0	0	<NA>	0	−1	2
103	0	0	<NA>	0	−1	1

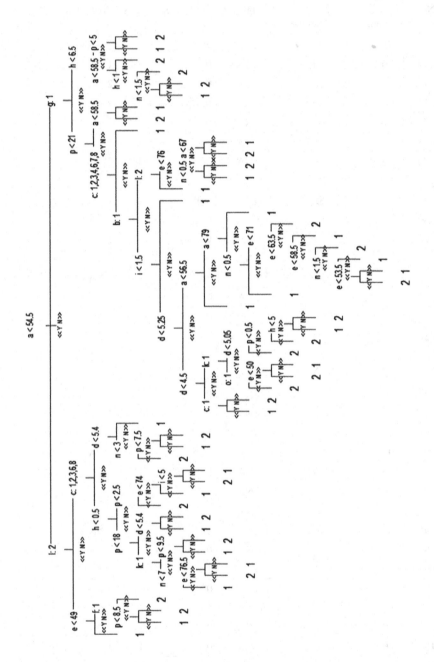

Fig. 3. Random forest tree

−1 represents leaf node. Last column is prediction that shows class label. If the node is non leaf node then this column contains value <NA> which means class identification is not required. Figure 3 is the tree representation of random forest generated rule.

7 Conclusion

Cataract condition develops in the lens of eye. Ophthalmologist consider numerous factors like living habits, age, gender as well medical conditions like diabetes, cholesterol level etc. for cause of cataract. In consultation with ophthalmologist, primary data was collected and studied using random forest algorithm. Random forest algorithm has given lowest OOB error estimation when value of set.seed was set to 3 and value of mtry set to 12. From Table 1, it is concluded that most important attribute for predicting the cataract is age and other factors in the order of importance has been shown in the Figs. 1 and 2. To predict the presence of cataract in the patient have been shown as rules in Table 6 and also visualized in Fig. 3. From confusion matrix shown in Table 3, it is concluded that for cataract yes the classification is more accurate (error is 0.2516556) and less accurate (0.3698630) for no cataract. Variable importance and tree path are useful to predict possibility of the cataract in individuals.

References

1. Anyanwu, M.N., Shiva, S.G.: Comparative analysis of serial decision tree classification algorithms. Int. J. Comput. Sci. Secur. (IJCSS) **3**(3), 230–240 (2009)
2. Sharma, H., Kumar, S.: A survey on decision tree algorithms of classification in data mining. Int. J. Sci. Res. (IJSR) **5**, 2094–2097 (2016)
3. Gupta, B., et al.: Analysis of various decision tree algorithms for classification in data mining. Int. J. Comput. Appl. **163**(8), 15–19 (2017). (0975–8887)
4. Kesavaraj, G., Sukumaran, S.: A study on classification techniques in data mining. IEEE-31661, 4–6 July 2013
5. Black, P.E.: Greedy Algorithm, in Dictionary of Algorithms and Data Structures, U.S. National Institute of Standards and Technology, February 2005. NIST-greedy algorithm
6. Alaoui, S.S., Labsiv, Y., Aksasse, B.: Classification algorithms in data mining. Int. J. Tomogr. Simul **31**, 34–44 (2018)
7. Breiman, L. Random Forests. Mach. Learn. **45**, 5–32 (2001). https://doi.org/10.1023/A:1010933404324
8. Goel, E., et al.: Int. J. Adv. Res. Comput. Sci. Softw. Eng. **7**(1), 251–257 (2017)
9. Brieman, L.: Random forests. Mach. Learn. **45**, 5–32 (2001)
10. Uwe, K., Dunemann, S.O.: SQL database primitives for decision tree classifiers. In: CIKM 2001. ACM, Atlanta (2001)
11. Denil, M., et al.: Narrowing the gap: random forests in theory and in practice. In: Proceedings of the 31st International Conference on Machine Learning, Beijing, China, vol. 32. JMLR: W&CP (2014)
12. Kaur, S., Grewal, A.K.: A review paper on data mining classification techniques for detection of lung cancer. Int. Res. J. Eng. Technol. (IRJET) **03**(11), 1334–1338 (2016). e-ISSN: 2395-0056, p-ISSN: 2395-0072
13. Parashar, H.J., Vijendra, S., Vasudeva, N.: An efficient classification approach for data mining. Int. J. Mach. Learn. Comput. **2**(4), 466 (2012)

14. Kumar, S.V.K., Kiruthika, P.: An overview of classification algorithm in data mining. Int. J. Adv. Res. Comput. Commun. Eng. **4**(12), 255–257 (2015). ISSN (Online) 2278-1021, ISSN (Print) 2319 5940

15. https://www.who.int/blindness/causes/priority/en/index1.html. Accessed 29 Sept 2019

16. Janitza, S., Hornung, R.: On the overestimation of random forest's out-of-bag error. PLoS ONE **13**(8), e0201904 (2018). https://doi.org/10.1371/journal.pone.0201904

17. Hastie, T., et al.: The Elements of Statistical Learning Data Mining, Inference, and Prediction, 2nd edn. Springer, New York (2009). https://doi.org/10.1007/978-0-387-84858-7

18. Nayer, A.M.: Detecting diabetes mellitus using machine learning. Int. J. Comput. Syst. **03**(12), 670–677 (2016). ISSN 2394-1065

19. Beaulac, C., Rosenthal, J.S.: Predicting university students academic success and major using random forest (2019). https://doi.org/10.1007/s11162-019-09546-y

20. Sugandhi, C., Yasodha, P., Kannan, M.: Analysis of a population of cataract patients databases in weka tool. Int. J. Sci. Eng. Res. **2**(10), 1 (2011). ISSN 2229-5518

21. Niya, C.P.: Automatic cataract detection and classification systems: a survey. TechS Vidya e-J. Res. **3**(2014–15), 28–36 (2015). ISSN 2322-0791

Yawn Detection for Driver's Drowsiness Prediction Using Bi-Directional LSTM with CNN Features

Sumeet Saurav[1,3](✉), Shubhad Mathur[2], Ishan Sang[2], Shyam Sunder Prasad[1,3], and Sanjay Singh[3]

[1] Academy of Scientific and Innovative Research (AcSIR), Ghaziabad, India
sumeet@ceeri.res.in
[2] Birla-Institute of Technology and Science (BITS, Pilani), Goa Campus, Goa, India
[3] CSIR - Central Electronics Engineering Research Institute (CSIR-CEERI), Pilani, India

Abstract. Drowsiness of drivers is a critical problem and has recently attracted a lot of attention from both academia and industry. A real-time driver's drowsiness detection system is often considered as a crucial component of an Advanced Driver Assistance System (ADAS). Although, there are a number of physical parameters associated with drowsiness like blink frequency, eye closure duration, pose, gaze, etc., yawing can also be used as an indicator of drowsiness. This work presents a novel deep learning-based framework for driver's drowsiness prediction based on yawn detection in a video stream. The proposed approach uses a combination of a convolutional neural network (CNN), 1D-CNN, and bi-directional LSTM (Bi-LSTM). In the first step, the pipeline extracts the mouth region from each frame of the video using a combination of face and landmark detector. In the subsequent step, spatial information from the mouth region is extracted using a pre-trained deep convolutional neural network (DCNN). Finally, temporal information which models the evaluation of yawn using the extracted mouth feature is learned using a blend of 1D-CNN and bi-directional LSTM (Bi-LSTM). Experiments were performed on manually extracted and annotated video clips obtained from two publically available drowsiness detection dataset namely YawDD and NTHU-DDD. Experimental results show the effectiveness of the proposed approach both in terms of recognition accuracy and computational efficiency. Thus, the proposed pipeline is a good candidate for real-time implementation of yawn detection system for driver's drowsiness prediction on an embedded device.

Keywords: Drowsiness detection · Convolutional neural networks · Long short-term memory (LSTM) · Bi-directional LSTM (Bi-LSTM)

1 Introduction

According to a recent article [1], in India more than 150, 000 people are killed every year due to road accidents which is much higher than the developed countries like US, where the count was 40,000 in 2016. Another report of the American National Highway Traffic Safety Administration (NHTSA) [2], states that there were 803 causalities which were

© Springer Nature Switzerland AG 2020
U. S. Tiwary and S. Chaudhury (Eds.): IHCI 2019, LNCS 11886, pp. 189–200, 2020.
https://doi.org/10.1007/978-3-030-44689-5_17

reported in 2016 due to drowsy driving. Therefore, drowsiness of drivers is considered as a critical problem and has attracted a lot of attention from both academia and industry. At present, many renowned multinational automobile companies like Nissan, Toyota, and Volkswagen are working in this area to create technologies which could mitigate the driver's drowsiness related issues by issuing a warning signal to the drivers.

There are a number of physiological parameters associated with drowsiness like eye closure duration, blink frequency, percent of times the eyes are close (PERCLOS), pose, gaze and nodding frequency which could be extracted using a visual sensor (Camera) [3]. Apart from these, another parameter which could also be a good indicator of driver's drowsiness is yawn. Yawning can be best defined as an involuntary act of gaping of mouth and deep inhalation followed by shallow exhalation. It is an important indicator to detect drowsiness and fatigue at an early stage.

During yawing, mouth opens wide and a change is observed in geometric features of mouth. Many approaches which use geometrical features extracted from mouth for yawn detection could be found in [4–6]. Various color-based approaches have also been proposed for mouth and lips detection [7]. In [8], the authors have used Gravity-Center template for face detection followed by mouth corner extraction using grey projection and Gabor wavelets. The authors in [9] segmented the image texture regions using stochastic region merging strategy, then skin color and texture models is used for classification. The work reported in [10] used two cameras to their advantage: a low-resolution camera for the face and a high-resolution one for the mouth. Haar-like features were then used to detect yawning based on the ratio of the height to width of mouth. Work presented in [11] used the static supervised classification based on log-polar signatures approach for open or closed mouth detection. Another work proposed in [12], used a combination of SVM based face detector, gradient edge detectors-based mouth region locator and circular Hough transform based yawn detector. Authors in [13] detected rate and amount of changes in mouth using back projection theory through a modified implementation of the Viola-Jones algorithm for face and mouth detection. Another method presented in [14] proposed to detect yawning using geometric and appearance features of mouth and eye regions and were also able to detect hand-covered and uncovered yawns. A combination of CNN and LSTM for yawn detection could be found in [15].

Our proposed method for yawn detection in some extent is similar to the work presented in [15]. However, there are a number of add-ons. First, in [15] the authors have used pre-trained CNN model trained on ImageNet, but in this work we pre-trained a deep convolutional neural network (DCNN) from scratch using FER+ database and used it for extracting spatial information from the mouth region. Secondly, the work in [15] have used stacked LSTM network for modelling temporal information related with yawn, but this work uses a combination of 1D-CNN and Bi-LSTM for the same. In our method, a camera installed on the dashboard/side mirror of the car continuously captures video streams. From these video streams, face is first detected which is then passed to a landmark detector. Using coordinates of the facial landmarks the mouth region is extracted. From the mouth region, deep features are extracted using a pre-trained DCNN. Finally, features extracted from 32 frames are fed to a combination of 1D-CNN and Bi-LSTM to detect a yawn.

The remainder of this paper is divided as follows: Sect. 2 presents an overview of our proposed method. Experiments results and discussions are dealt in Sect. 3. We finally conclude in Sect. 4.

2 Proposed Method

The proposed framework used for detection of yawn in a video stream is shown in Fig. 1.

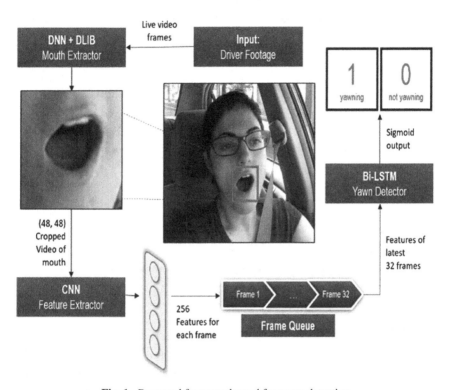

Fig. 1. Proposed framework used for yawn detection

As could be seen from the above figure, the framework consists of three main steps: mouth region extractor, deep feature extractor, and yawn detector. Each of these steps are briefly discussed below.

2.1 Mouth Region Extractor

In order to extract the mouth region from a given video stream, the mouth region extractor needs to detect the face region and then locate and track facial landmarks on the detected face. We used OpenCV's DNN model (ResNet-10 architecture) to detect the face region and Dlib's shape predictor [16] (68-face-landmarks) to get the facial landmarks. The points corresponding to the nose (33), lip corners (48 and 54) and near the mouth (48 to 67) (as shown in Fig. 2) are then used to find the mouth region. The model finds the center of the mouth region using the average of all the points near the mouth. Then using the points 33, 48 and 54 it estimates the size of the mouth region and creates a square bounding box around it. This box is then resized to size (48, 48) and passed on to the feature extractor.

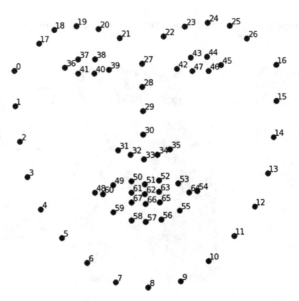

Fig. 2. Set of 68-points landmark locations available in [16]

2.2 Deep Feature Extractor

To extract spatial information from the extracted mouth region a deep convolutional neural network (DCNN) has been used in this work. The proposed DCNN was trained from scratch using the FER+ database [17]. Details of the DCNN architecture has been shown in Fig. 3.

The network takes as input a 48 × 48 pixels grayscale image (one single frame of the mouth video). There are four convolutional layers, three max pooling layers, two dense layers, and finally a SoftMax classifier layer. Once the DCNN is trained, it was used for extracting features from the mouth images. The dimension of the features extracted using the network is 256.

2.3 Yawn Detector

A typical yawn is a dynamic and continuous process that lasts for around 2–3 s and has 3 phases: the person opening her/his mouth, the prolonged open state of the mouth and, then its closing. To find a relationship between the features of consecutive frames and determine if the person is yawning, a recurrent neural network (RNN) is used.

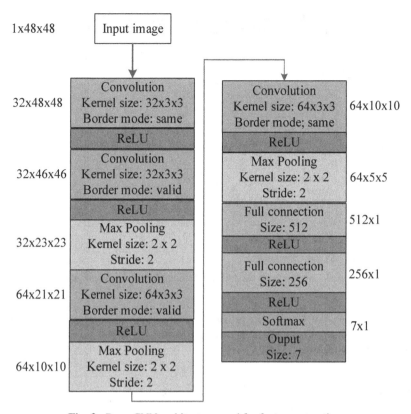

Fig. 3. Deep CNN architecture used for feature extraction

RNNs use the features of the frame (it is currently processing) and also, the output for the previous frame to generate prediction for each frame. However, a vanilla RNN suffers from vanishing and exploding gradients problems and also struggles to learn long term dependencies. To counter these, special type of RNNs, Long Short- Term Memory Networks (LSTMs) are used instead [18]. LSTMs use memory cells to store information and input & forget gates to control its flow, which enables them to discover long range temporal relationships.

In our system, we used a combination of 1D-CNN and Bi-directional LSTM as shown in Fig. 4. The principle of a Bi-directional RNN is to split the neurons of a regular RNN into two parts: one for forward pass in the positive direction and another for the negative direction. This enables the neuron, processing the current frame, to gather information from both the past and the future frames, which leads to more robust learning.

1D-CNN layer (1 * 256, same padding) convolves on the 32 frames. The yawn detector takes as input 32 frames at a time and performs forward and backward propagation on them. For inference, the model maintains a queue of the latest 32 frames of the video feed and passes them into the yawn detector. It then takes the prediction for the latest frame (sigmoid output) as the output for the latest frame of the video. We kept a threshold value (A) for the sigmoid output and another threshold (B) for the number of continuous

Layer (type)	Output shape	Parameters #
Input	(32, 256)	0
Conv 1D	(32, 256)	196864
Activation (ReLU) + BN + Dropout	(32, 256)	1024
Bi LSTM	(32, 256)	394240
Dense	(32, 256)	65792
Activation (ReLU) + BN + Dropout	(32, 256)	1024
Dense	(32, 1)	257
BN + Activation (Sigmoid)	(32, 1)	4

BN: Batch Normalization

Fig. 4. Architecture used for modelling temporal information

frames in which 'YAWNING' was predicted. If the sigmoid output is beyond A and the number of continuous 'YAWNING' frames is beyond B, the frame is classified as 'YAWNING', else as 'NOT YAWNING'.

3   Experimental Results and Discussions

This section gives details of various datasets and experiments which were performed in this study.

3.1   Dataset Preparation

Three different datasets namely FER+ , YawDD, and NTHU-DDD were used in this study. The FER+ [17] dataset is the re-tagged version of the FER2013 database consisting of facial images of different expressions (happy, neutral, sad, surprise, disgust, contempt, anger, and fear). The dataset comprises of 35,541 facial images of resolution 48 × 48 pixels divided into training, validation, and test set. We used this database to pre-train the DCNN wherein only facial images from 7 expressions (anger, disgust, happy, fear, neutral, sad, and surprise) were considered and the training samples were created by merging the actual training and test set provided in the dataset. The validation set provided in the dataset was used as validation set during training. Sample images corresponding to different expressions from the dataset has been shown in Fig. 5.

The YawDD dataset [19] is an open-source dataset comprising of videos of drivers with various facial characteristics and targeted towards designing and testing algorithms and models for yawn detection. Two different variants of the database are available, one captured using a camera mounted on the dash board of the car and the other captured using camera mounted on the side mirror of the car. We used the later variant of the

Fig. 5. Sample images from FER+ 7 expression dataset (left to right: anger, disgust, fear, happy, neutral, sad, and surprise)

database which contains a total of 322 videos of male and female drivers of different ethnicities with/without glasses/sunglasses. The videos have been labelled into three classes: no talking (normal), talking, and normal. The videos have been captured using a RGB camera at 30 frames per second with a resolution of 640 × 480 pixels. Some of the sample frames extracted from the video clips of the database has corresponding to normal, talking, and yawning classes have been shown in Figs. 6, 7 and 8 respectively.

Fig. 6. Sample frames from the normal class video clip (YawDD dataset)

Fig. 7. Sample frames from the talking class video clip (YawDD dataset)

Fig. 8. Sample frames from the yawning class video clip (YawDD dataset)

The final database used is the NTHU-DDD database [20]. Videos in the dataset is divided into two splits: training and validation. The training dataset consists of 356 videos containing 18 subjects of different ethnicities recorded with and without wearing glasses/sunglasses under a variety of simulated driving scenarios, including normal driving, yawning, slow blink rate, falling asleep, burst out laughing etc., under day and night illumination conditions. The videos of the dataset were labelled either as sleepy/nonsleepy. However, in our experiments we manually extracted clips from the videos where the subject was either yawning or normal. Sample images extracted from the yawn video clip from the NTHU-DDD database has been shown in Fig. 9.

Fig. 9. Sample frames from the yawning class video clip (NTHU-DDD dataset)

Unfortunately, both the datasets (YawDD and NTHU-DDD) contained video clips which were labelled as yawning/sleepy, but much of the video content in them covered no yawning at all. Thus, it was necessary to crop out the yawning parts of the videos (a typical yawn lasting 3–5 s), otherwise the model would have wrongly learnt to classify people who were acting normal/talking as yawning. We also added the rest of the non-yawning parts of these 'yawning' videos in our 'non yawning' set. After cropping, we removed a few videos in which the participants' yawn appeared to be extremely fake. Finally, the total number of yawning clips obtained was 205 and 526 from YawDD and NTHU-DDD dataset respectively. The number of normal/talking clips obtained from these were 318 from YawDD and 178 from NTHU-DDD datasets respectively.

3.2 Model Training and Results

Training of the models was done on a GPU machine having Ubuntu 16.04 OS, 32 GB RAM, and NVIDIA GeForce 1080 GPU with 8 GB memory. We used Keras [21] with Tensorflow backend for conducting the experiments.

The feature extractor deep convolutional neural network (DCNN) was trained on the popular FER+ dataset. Different on-the-fly data augmentation techniques (horizontal flip, random rotation within the range of 10°, random zooming by x0.1, and normalization) were used to avoid overfitting. These augmentations were performed using the in-built Keras ImageDataGenerator utility. The model was trained using a batch size of 64. The learning rate was kept fixed to 1e-5 and the Adam optimizer was used with categorical cross-entropy loss function. After training the model for 200 epochs, we got a validation accuracy of around 75%. The learning curve obtained after training the DCNN is shown in Fig. 10. Although we could further improve the validation accuracy, but we did not attempt, as the trained network with 75% validation accuracy was found sufficient enough to extract discriminative features from the extracted mouth region.

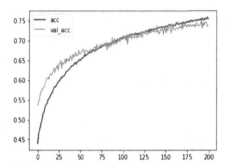

Fig. 10. Deep convolutional neural network learning curve (x-axis: epochs, y-axis: accuracy in %)

Once the feature extractor model got trained, we used it to extract features from each frame of the video clips and saved it on the disk. From all the videos in our dataset (YawDD and NTHU-DDD), we got a total of 1114924 non-yawning frames and 21919 yawning frames.

To train the yawn detector model the extracted saved features were loaded from the disk. We then used a sliding window approach on this consecutive array of frames to create numerous windows having frame size of 32. This array of windows was then randomly shuffled, and split into training and validation sets (70:30 ratio). After doing this, we ended up with 400227 windows (samples) for validation and 933860 windows (samples) for training. Each window had 32 frames and each frame had a 256-length feature vector. We trained this model with a learning rate of 1e-5, using Adam optimizer and a binary cross entropy loss function. We took a batch size of 64 and shuffled the dataset before each epoch. Batch Normalization and Dropout (with a rate of 0.4) is

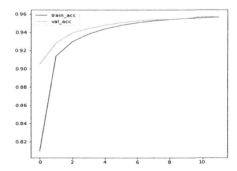

Fig. 11. Yawn detector learning curve (x-axis: epochs, y-axis- accuracy in %)

applied after every layer. We also used kernel and recurrent (L2) regularizes to prevent the model from overfitting. Our model achieved a final validation accuracy of 96.48% and training accuracy of 95.66% after being trained for 12 epochs as could be seen from the Fig. 11.

Once trained the model was tested on a laptop having 8 GB RAM, Intel i7 processor core, Ubuntu 18.04.2 LTS OS and Keras 2.2.4 deep learning library with TensorFlow backend. The inference speed achieved by the proposed yawn detector was ~25 FPS (frames per second) achieving real-time performance. Sample detection results obtained using our proposed framework for yawn detection are shown in Fig. 12.

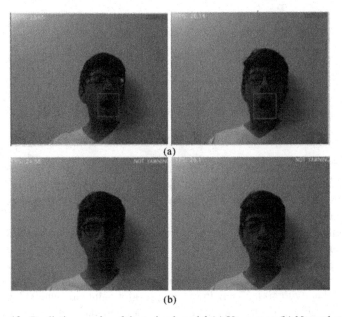

(a)

(b)

Fig. 12. Prediction results of the trained model (a) Yawn case (b) Normal case

From the detection results, one can find that the proposed approach is quite robust towards lighting variations and is able to differentiate between different mouth states properly.

4   Conclusion

In this work, we presented an efficient deep learning-based approach for yawn detection targeted towards driver's drowsiness prediction. The proposed algorithmic pipeline consists of face detector, a landmark detector, a deep convolutional neural network (DCNN), and a combination of 1D-CNN and bi-directional LSTM (Bi-LSTM). Face and facial landmarks detector have been used for extracting the mouth region from the video frames. The DCNN pre-trained on FER+ database has been employed to extract the spatial information from the extracted mouth regions. Finally, a blend of 1D-CNN and Bi-LSTM has been used to model the temporal information associated with a yawn sequence. Experiments were done on a manually annotated yawing and normal clips cropped from the YawDD and NTHU-DDD datasets. From the experimental results, we found the effectiveness of the proposed approach both in terms of recognition accuracy and computational efficiency. Thus, the proposed pipeline is a good candidate for real-time implementation of a yawn detector on an embedded platform for driver's drowsiness prediction.

References

1. https://timesofindia.indiatimes.com/india/over-1-51-lakh-died-in-road-accidents-last-year-up-tops-among-states/articleshow/72078508.cms. Accessed 18 Nov 2019
2. National Center for Statistics and Analysis: 2016 fatal motor vehicle crashes: Overview. (Traffic Safety Facts Research Note. Report No. DOT HS 812 456) (October 2017). National Highway Traffic Safety Administration, Washington, DC
3. Sikander, G., Anwar, S.: Driver fatigue detection systems: a review. IEEE Trans. Intell. Transp. Syst. **20**(6), 2339–2352 (2018)
4. Ji, Q., Zhu, Z., Lan, P.: Real-time nonintrusive monitoring and prediction of driver fatigue. IEEE Trans. Veh. Technol. **53**(4), 1052–1068 (2004)
5. Wang, T., Shi, P.: Yawning detection for determining driver drowsiness. In: Proceedings of IEEE International Workshop on VLSI Design and Video Technology, May 2005, pp. 373–376 (2005)
6. Rongben, W., Lie, G., Bingliang, T., Lisheng, J.: Monitoring mouth movement for driver fatigue or distraction with one camera. In: Proceedings of the 7th IEEE International Conference on Intelligent Transportation Systems, October 2004, pp. 314–319 (2004)
7. Lu, Y., Wang, Z.: Detecting driver yawning in successive images. In: 1st International Conference on Bioinformatics and Biomedical Engineering, July 2007, pp. 581–583 (2007)
8. Fan, X., Yin, B.C., Sun, Y.F.: Yawning detection for monitoring driver fatigue. In: IEEE International Conference on Machine Learning and Cybernetics, August 2007, vol. 2, pp. 664–668 (2007)
9. Medeiros, R.S., Scharcanski, J., Wong, A.: Multi-scale stochastic color texture models for skin region segmentation and gesture detection. In: IEEE International Conference on Multimedia and Expo Workshops (ICMEW), July 2013, pp. 1–4 (2013)
10. Li, L., Chen, Y., Li, Z.: Yawning detection for monitoring driver fatigue based on two cameras. In: 12th IEEE International Conference on Intelligent Transportation Systems, October 2009, pp. 1–6 (2009)
11. Bouvier, C., Benoit, A., Caplier, A., Coulon, P.-Y.: Open or closed mouth state detection: static supervised classification based on log-polar signature. In: Blanc-Talon, J., Bourennane, S., Philips, W., Popescu, D., Scheunders, P. (eds.) ACIVS 2008. LNCS, vol. 5259, pp. 1093–1102. Springer, Heidelberg (2008). https://doi.org/10.1007/978-3-540-88458-3_99
12. Alioua, N., Amine, A., Rziza, M.: Driver's fatigue detection based on yawning extraction. Int. J. Veh. Technol. **2014** (2014)
13. Omidyeganeh, M., et al.: Yawning detection using embedded smart cameras. IEEE Trans. Instrum. Meas. **65**(3), 570–582 (2016)
14. Jie, Z., Mahmoud, M., Stafford-Fraser, Q., Robinson, P., Dias, E., Skrypchuk, L.: Analysis of yawning behaviour in spontaneous expressions of drowsy drivers. In: 13th IEEE International Conference on Automatic Face & Gesture Recognition, FG 2018, May 2018, pp. 571–576 (2018)

15. Zhang, W., Su, J.: Driver yawning detection based on long short-term memory networks. In: IEEE Symposium Series on Computational Intelligence (SSCI), November 2017, pp. 1–5 (2017)
16. King, D.E.: Dlib-ml: a machine learning toolkit. J. Mach. Learn. Res. **10**, 1755–1758 (2009)
17. Barsoum, E., Zhang, C., Ferrer, C.C., Zhang, Z.: Training deep networks for facial expression recognition with crowd-sourced label distribution. In: Proceedings of the 18th ACM International Conference on Multimodal Interaction, October 2016, pp. 279–283. ACM (2016)
18. Gers, F.A., Schmidhuber, J., Cummins, F.: Learning to forget: continual prediction with LSTM (1999)
19. Abtahi, S., Omidyeganeh, M., Shirmohammadi, S., Hariri, B.: YawDD: a yawning detection dataset. In: Proceedings of the 5th ACM Multimedia Systems Conference, March 2014, pp. 24–28. ACM (2014)
20. Weng, C.-H., Lai, Y.-H., Lai, S.-H.: Driver drowsiness detection via a hierarchical temporal deep belief network. In: Chen, C.-S., Lu, J., Ma, K.-K. (eds.) ACCV 2016. LNCS, vol. 10118, pp. 117–133. Springer, Cham (2017). https://doi.org/10.1007/978-3-319-54526-4_9
21. Chollet, F.: Keras (2017). https://github.com/fchollet/keras

Assistive Living and Rehabilitation

Robotic Intervention for Elderly - A Rehabilitation Aid for Better Living

Richa Pandey[✉] and Mainak Mandal

Mechanical Engineering Department, BIT Mesra, Ranchi, India
richarp@rediffmail.com, be10449.17@bitmesra.ac.in

Abstract. This paper concentrates on new technology development to support human mobility of ageing people. The development is based on human biomechanics. Movement of different joints and combination of joints are observed closely. On that basis, attempt has been made to design exoskeleton to support different body parts of old person. This research will help aged people for normal movement with ease. The paper aims to the major objectives which includes 1. To develop an exoskeleton powerful enough to transmit required high force to the muscles of the patients, so that patients don't feel any difference between human physiotherapists & the exoskeleton. 2. To make the process of physiotherapy exercise beneficial for stroke patients. The proposed motor actuated wearable mechanical structure for upper limb (also known as upper limb exoskeleton) is to be integrated with virtual reality games. The elderly persons are noticed to have major problem in the lower limb i.e. hip, knee and ankle. The Paper introduces a measure to aid the people with lower limb disability.

Keywords: Biomechanics · Rehabilitation · CAD model

1 Introduction

The main idea is to apply the technology of exoskeleton and virtual reality in the field of healthcare for neuro rehabilitation/physiotherapy of stroke patients. We have used a motor powered mechanical device (exoskeleton) which is wearable in the upper limb. After wearing this device in the arms, all the joint rotation of the user can be controlled by this device. Rotary as well as linear motors have been used to give joint rotation to the user's arms. This device has been provided with 5 degrees of freedom, providing three types of rotation in shoulder joint and two types in elbow joint. There will be sensors (mainly EMG sensor or force sensor) which will detect the small effort by the user to rotate one of his joints. That small effort will be amplified with the help of motors and joint of the patient will be rotated due to this amplification. In this way, physical exercise of the user will be done. Other types of sensors, the rotary encoders which will detect the angle of rotation of the joints and will send the data to the microcontroller for closed loop control of the motors for accuracy in arm movement.

© Springer Nature Switzerland AG 2020
U. S. Tiwary and S. Chaudhury (Eds.): IHCI 2019, LNCS 11886, pp. 203–211, 2020.
https://doi.org/10.1007/978-3-030-44689-5_18

2 Upper Limb Anatomy

Human upper limb consists mainly of the three joints at shoulder, elbow joint and joint at and after wrist comprising of finger joints. Fingers, which are important parts of hand of every human being consisted of several joints in themselves. Figure 1 shows that three bones are mainly there at the shoulder complex: the clavicle, scapula and humerus, and four articulations: the glenohumeral, acromioclavular, strenoclavicular and scapulothoric, with the thorax as stable base. Mostly, what we say in daily life as the shoulder joint is mainly glenohumeral joint.

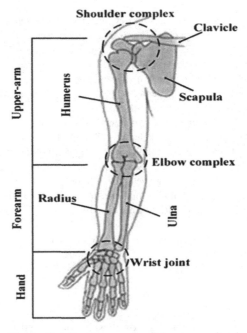

Fig. 1. Details of bones present in a human hand.

The whole of the upper limb is connected to the main backbone at only one joint. This joint is referred as Sternoclavicular joint. The acromioclavicular joint connects the lateral end of the clavicle and acromion of the scapula. The sternoclavicular joint is a compound joint containing two compartments in itself separated by the articular disks. It is formed mainly by the parts of clavicle, sternum and cartilage of the first rib. In a real sense the scapulothoracic joint cannot be considered as a joint as it is a bone-muscle-bone articulation which is not synovial. It is formed actually by the female surface of the thorax. It is considered as a joint only when describing motion of the scapular over the thorax. Shoulder complex is actually described as a ball and socket joint. It is formed mainly by the humerus and female part of the scapula. However, position of the center of rotation of shoulder joint is changing with upper arm motions in the lateral plane. Main motions provided by the other joints of shoulder complex and glenohumeral joint of the shoulder complex are the shoulder flexion-extension, shoulder abduction/adduction and internal-external rotation. The elbow complex in particular includes the elbow joint and the radioulnar joints. It is actually a compound joint consisting of two other joints: the humeroradial between the capitulum and radial head, and the humeroulnar between the trochlea and the trochlear notch of the ulnar. The humeroradial joint is a ball and socket joint. However, its close association with humeroulnar and superior radioulnar joint motion from three to two DOF (Degrees Of Freedom). As a whole, the elbow joint complex allows two DOF, flexion/extension and supination/pronation. The whole mechanics is shown in the Fig. 2 (a) and (b).

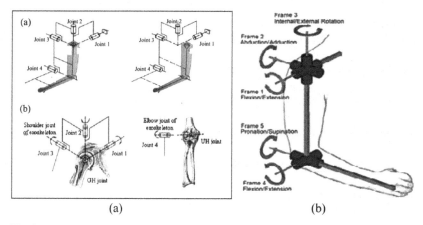

Fig. 2. (a) and (b) shows the idea of joints in human hand and their degree of freedom.

These two figures above (Fig. 2 a and b) shows three different joint locations and two other joints in elbow. The shoulder can do all the three rotations i.e. roll (in X axis), pitch (in Y axis) and yaw (in Z axis) and the elbow can do only two rotations i.e. roll and pitch.

3   Lower Limb Anatomy

During the human mobility, the three lower-body joints in each leg need to be considered; the main characteristics and functionalities are as follows:

- Hip joint: It is the largest ball-and-socket joint in humans which supports the weight of the human body during static and dynamic motions.
- Knee joint: This single degree of freedom joint having complex mechanisms that have evolved over time to bear high weight and pressure loads while providing flexible movements especially while performing daily motions.
- Ankle joint: This complex hinge joint used mainly in balance of the body structure and providing the required driving force during walking.

Figure 3 shows the internal articulation mechanism of a human hip joint which is also known as an acetabulofemoral joint. Because it is formed by the deep ball and socket configuration of the acetabulum of the pelvis and the head of the femur. It provides balance to the whole body structure and maintain a desired pelvis inclination angle to provide greater all-round stability as shown in Fig. 3 (see for example, Towson Orthopaedics [1]) with its 3 rotary degrees of freedom (DOF). There are anatomical differences between male and female hip joints; female hip joints have greater hip angles for childbirth but this increases the risk of torsional knee injuries for female (see Lumen Learning [2]). Hence females are likely to need more assistance then males. The knee has a complex 1 DOF joint important for human locomotion and most humans face difficulties with their knees as they aged; most of the people in India suffer from knee problems before they are 65 years old (as reported in India Today [3]). The different knee complexities arise due to the need to provide two distinct joints from the motion due to the tibiofemoral joint and the force transfer through the patellofemoral joint as knee motions are performed (see Lavengie and Norkin [4]). The two motions are shown in Fig. 2 (modified diagram from Leiden University Medical Centre [5]). Figure 4 shows different joints in knee.

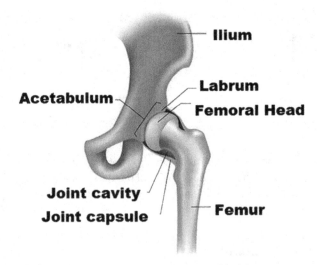

Fig. 3. Bone details of a human hip joint

The tibiofemoral joint consists of a double condyloid joint in which the femur has a rolling and sliding motion over the tibia to give the knee motion and the patellofemoral joint involves motion of the patella so that the required force if transferred as needed for the motion. Here in the designing of the exoskeleton for the knee having the tibiofemoral articulation has the most important motion discussed here. This joint made of a double condylar (medial and lateral condyles as shown in Fig. 5).

This condyloid joint is an ovoid articular surface. This joint mainly moves in two planes, while allowing flexion and extension. The knee joint (or double condylar) has two condylar surfaces with two sliding surfaces on the same side with a gap between them on the femur (or femoral condyles) and tibia (or tibial plateau). These details are

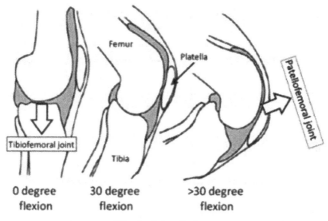

Fig. 4. Joints in knee

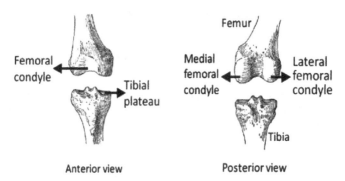

Fig. 5. Views of the condyle and plateau.

shown in Fig. 5. At the time when the leg muscles apply the force to make a required motion then the femoral condyles roll inside the tibial plateau and the excess rotation is restricted by the ligaments. Due to the constraints of the knee ligaments, there is no pure rolling at the knee joint but a mixture of rolling and sliding of the femoral condyles over the tibial plateaus.

4   Mechanical Design

The mechanism of the exoskeleton is slightly different from the actual bone mechanism. The fundamental aim of the mechanism is to perform as a parallel manipulator with the human arm, so that the necessary power can be transferred from the parallel manipulator to the human arm. This rig has a total of four degree of freedom. One of them on sagittal plane is active and other three is passive.

This mechanism is made in a way such that for a certain angle of rotation at gleno-humeral joint, there is some rotation at both glenohumeral and sternoclavicular joint. The proportionality of angle depends on the length PQ of the linkage PQR shown in

Fig. 6. This mechanism is mostly suitable since it has provision for rotation of collarbone too. But since the angle of rotation of collarbone varies from person to person, length of PQ also needs to be varied accordingly for best fit. The given mechanism is made for clavicle rotation of up to 24°. So, if people having less clavicle rotation can use it, a mechanical stopper will stop any further rotation of their clavicle bone. Here in this mechanism some linear actuators are used to actuate the humerus adjustment link and the active shoulder joint. A CAD model is depicted in Fig. 7.

Details:
a. Arm holder and slider:
The arm holder provides support and motion to the arm. It provides a passive degree of freedom providing the unique supination or pronation motions.

b. Elbow joint:
Joint 3 allows the flexion and extension movements while the shoulder joint gives abduction or adduction motions at the shoulder complex. In a human arm, the gleno-humeral joint moves independent of the shoulder complex up to almost 90° and the shoulder complex moves as a single unit. To control this, the shoulder joint was provided with a screw barrier that restricts its independent motion at certain required angle and the force of the linear actuator gets applied to the entire shoulder complex. Linear slider i.e. second slider part at arm allows to adjust the position of the arm holder by permitting different users to use it, since arm length (i.e. the length of humerous bone) of different users can be varied. Also, it helps in the slight variation of arm holder position during abduction and adduction movement. Analysis of the dynamics of the mechanism is done in the Adams software. One moment of the analysis is pictured below in Fig. 8. All the graphs in it are plotted against time.

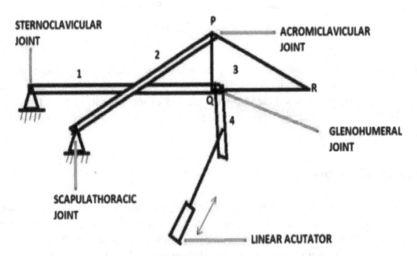

Fig. 6. Approximate mechanism for shoulder complex.

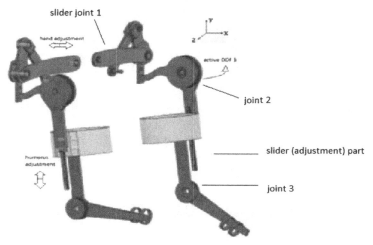

Fig. 7. The CAD model was designed using Autodesk Fusion 360 because of its flexible workbench options and modelling freedom.

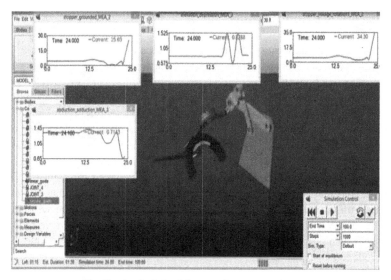

Fig. 8. Analysis in Adams software at a glance.

This exoskeleton is specially made for upper limb (Fig. 9) but if we add a foot rest at the end (i.e. at elbow part) then it can also be used for the lower limb.

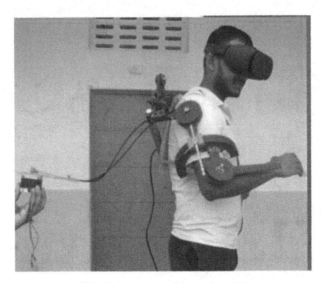

Fig. 9. Prototype of the cad model

5 Technological Innovations

This mechanism is actually made for the rehabilitation of the hand movement. Adding some ideas would make it even better. Such as, we can integrate this exoskeleton with VR system. We can touch virtual objects and track our hand motion in VR. The range of our hand motion can be changed manually. Motion capturing and force feedback will be also there in the device. Using various stiffness the exoskeleton applies precise motor control to distinguish the objects and applies variable force to simulate the difference between hard and soft objects. This system can be used for gaming as well as military, training field. For integration with games the software with virtual reality games controlled by rotary encoder data will be used. The game will be displayed through a USB cable on a mobile and that will be fed inside a virtual reality headset, so that whatever will be displayed on the pc/laptop will also be shown on VR headset. User will put on the exoskeleton and VR headset. Based on the game, he will just try to move his arm and with the help of motors in exoskeleton his arm will be rotated correspondingly (using rotary encoders).

While designing a mechanism to fit the human body, it is important to understand the human body parts and their bio-mechanics to give a comfortable feeling. Both lower and upper limb joints are used to carry out mobility and daily tasks as necessary. For example, during walking the hip, knee and the ankle joints provide 42%, 42%, 16% power respectively (Montegomery and Grabowski [6]).

Control is equally important to ensure the Exo-skeleton is wearable or not or will it be effective in providing support during any process. For this multiple sensor-actuator loops are planned to be studied consisting joint angle, translational or rotational joint velocities and accelerations and forces at joints so that a robust machine-user interface can be developed . Force sensors will automatically detect the movement what any user wants to do and accordingly the motion will be detected and the motors will drive the mechanism to replicate the motion exactly. This result will be easier to carry out to the user.

6 Conclusion

The paper is an attempt to present robotics as an aid and solution to the elderly person for happy living. The work encompasses the development of an assistive physiotherapy Aid which shall replace the human intervention and make the work simple and feasible. The major advantage of these aids is the user friendliness and better responsive in nature. The wearable can be worn any time and is comfortable and effective. The rehabilitation is the key factor of the proposed work and caters to the best of the needs of the user.

References

1. Towson Orthopaedic Associates website. www.towsonortho.com/specialties/joint-preservation-center/hip-joint-preservation. Accessed 28 Feb 2019
2. Lumen Learning. https://courses.lumenlearning.com/boundless-ap/chapter/the-hip/. Accessed 29 Feb 2019
3. India Today. www.indiatoday.in/lifestyle/wellness/story/ninety-percent-of-people-suffer-from-damaged-knee-joint-problem-by-the-age-of-60-to-65-years-279575-2015-12-30. Accessed 28 Feb 2019
4. Lavengie, P.K., Norkin, C.C.: Joint Structure and Function: A Comprehensive Analysis, 4th edn. FA Davis Company, Philadelphia (2005)
5. StartRadiology, Leiden Univ Medical Center website. www.startradiology.com/internships/orthopedics/knee/x-knee/. Accessed 28 Feb 2019
6. Montogomery, J.R., Grabowski, A.M.: The contributions of ankle, knee and hip joint work to individual leg work change during uphill and downhill walking over a range of speeds. R. Soc. Open Sci. 5(8), 180550 (2018)

ccaROS: A ROS Node for Cognitive Collaborative Architecture for an Intelligent Wheelchair

Mohammad Arif Khan[2], Sumant Pushp[1(✉)], and Shyamanta M. Hazarika[3]

[1] Human Centered Computing Lab, Central University of Jharkhand, Ranchi, India
sumantpushp@gmail.com
[2] Biomimetic and Cognitive Robotics Lab, Tezpur University, Tezpur, Assam, India
[3] Indian Institute of Technology, Guwahati, India

Abstract. For effective Human-Robot Interaction (HRI), an intelligent wheelchair (IW) need to be cognitively enhanced. Robot Operating System (ROS) has been steadily gaining popularity among robotics researchers as an open source framework for robot control. This paper presents ccaROS - a new ROS node for a Cognitive Collaborative Architecture to achieve better HRI for an IW. The design of the ROS node is presented. It provides mechanisms for obstacle avoidance, detection and adaption of user's navigational strategy; seamless switching of driving control from machine to human and vice-versa. This would not only assist to achieve safe navigation but also allow retention of residual skills of the user. The effectiveness of ccaROS has been evaluated through simulation studies within a ROS-USARSim environment.

Keywords: Intelligent wheelchair · ROS node · Cognitive Collaborative Architecture

1 Introduction

For people who suffer from mobility impairment, intelligent wheelchairs (IWs) are a sought-after solution. However, users of IWs often struggle to drive safely and effectively; resulting in the loss of their residual skills. Several prototypes have been developed and control algorithms proposed to assist users drive safely [4,7]. Nevertheless, more often than not, the user is relegated to being a *rider* rather than taking advantage of the user's potential; exceptions being [3,12]. IWs must interact with the human as a team-mate! Cognitive embodiment through cognitive architectures enable better Human Robot Interaction (HRI) [6,10]. For retention of residual skills, the user must be provided assistance-as-required; cognitive architecture has been proposed to provide help-if-needed [8].

Cognitive Collaborative Architecture (CCA) [8] is inspired from [10,11] and is an adaption of ACT-R [1]. Figure 1 shows the CCA. The architecture incorporates three layers: a. user interface layer - provide the interface between user

© Springer Nature Switzerland AG 2020
U. S. Tiwary and S. Chaudhury (Eds.): IHCI 2019, LNCS 11886, pp. 212–221, 2020.
https://doi.org/10.1007/978-3-030-44689-5_19

and machine, where the user can communicate with the system through the conventional interface (e.g. keyboard or joystick), b. local control layer - to interact with the low-level electronics - sensors and actuators and c. superior control layer to perform heuristic computation responsible for driving in two different modes - manual control and automated control.

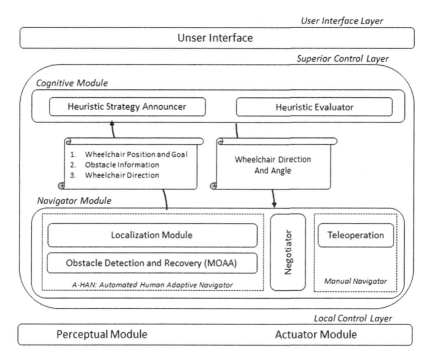

Fig. 1. The Cognitive Collaborative Architecture (CCA), from [8]. The CCA is arranged in three layers: a. User Interface Layer, b. Superior Control Layer and c. Local Control Layer.

Spatial representation is ubiquitous within robot planning and navigation. However, understanding spatial structure is complex. Qualitative representation seems closer to how human comprehend space. Inline with the above, embodying human-like navigation strategy in CCA lead to more effective way for an agent to encode the behaviour of its human user. This underlies the fact that a robotic agent could be a better teammate if it would observe and follow a good decision taken by human and take initiative to override a decision otherwise.

Robot Operating System (ROS) [9] is an open-source, meta-operating system. ROS provides the services one would expect from an operating system, including hardware abstraction, low-level device control. However, it cannot be seen as a traditional operating system; ROS is a collection of APIs, tools, and algorithms for effective and easy development and operation of robots. We present ccaROS - an implementation of the CCA within ROS. This is in line with published ROS node [2] designed to seamlessly interface between ROS and USARSim. ccaROS has been developed for effective HRI.

2   CcaROS: A ROS Node for CCA

ccaROS is an implementation of Cognitive Collaborative Architecture on top of ROS as a middleware. ccaROS include a number of predefined nodes to reduce the complexity of implementation by utilizing basic functionality to communicate with sensors and actuators. For example, localization is through AMCL routines. The proposed ROS Node is organized as a collection of packages to provide the opportunity for modular organization and scalability.

The *User Interface Layers* includes the input control methods required to drive the wheelchair by a human user. Whereas the *Local Control Layers* need to receive sensor data from a variety of sensors and to send a motor command to actuators. The implementation of these two layers does not bother much as there are numerous predefined packages available with ROS which can fulfill the requirement as expected. *Superior Control Layer* has close interaction with perceptual components and actuators. It is responsible for driving in manual and automated mode. Figure 2 shows the components of the superior control layer. *Negotiator* works as a control switch and alternates the control between *manual* and *automated control.* Automated controller possess obstacle avoidance mechanism and localization information through *MOAA* (Margin based Obstacle Avoidance Algorithm) [7] and *amcl* (Automated Monte Carlo localization - a predefined ROS node) respectively.

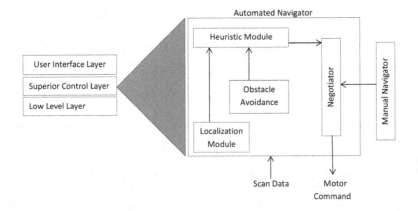

Fig. 2. Components of Superior Control Layer

2.1   Autonomous Control Within CCA

CCA is designed to provide assistance on demand and facilitate seamless switching. It does not look for the shortest path between two points, rather it assists the user in a way he desires to navigate. The requirement for *avoiding obstacle* is fairly simple and reasonable. A fix width margin is taken about the obstacles; moving towards the margin, alerts the system through an expectation of crash.

Figure 3 shows three decision states of MOAA. Assistance is not provided in a SAFE state. The moment IW approach the margin, control transfer between user and agent will take place for a short period (till the state become SAFE). On repeated mistake, state will become UnSAFE and automated agent will drive.

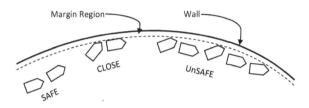

Fig. 3. Three States of MOAA

MOAA in Algorithm 1, is a fairly simple procedure for avoiding obstacles. However, in cooperation with the cognitive module, MOAA is powerful as it not only takes care of spatial detail but also looks for the temporal event. For instance, if the user commits mistakes, the automated agent every time overrides the control and moves to a safe zone. If the user does so repeatedly over a short period of time, numerous corrections will lead to jerk. MOAA posses a temporal aware parameter which tells if the user commits mistakes in the subsequent interval of time, it then override the manual control and drive the wheelchair in automated mode. The intuition is not only to overcome the jerk but also to reduce frustration due to unmanageable maneuver by a physically or cognitively impaired user.

To facilitate a proactive adaption, the heuristic module continuously monitors the navigation strategy followed by a user. The moment crash is predicted, it returns new orientation information based on the previous navigation strategy by the user. Here it is interesting to observe that whenever a crash occurs or is expected to occur, a human user will also try not crash. Therefore he/she may look for orientation recovery. Users of IW are cognitively or physically impaired, they may not succeed in most of their attempts. This may reduce their confidence and finally reduce motivation towards independent mobility. There could be many strategies a human user can opt during orientation recovery. We encoded three types of such strategies to be the part of CCA: 1. Following the wall 2. Towards center point and 3. Towards Goal.

Whenever crash occurs it is expected that negotiator should transfer the control from user to automated control (assuming that user of IW will fail to do the same by their own most of the time). At this moment automated controller receives new orientation information, which is based on the navigation strategy followed by the user in a few previous intervals. Therefore new orientation and relocation position will seem to be very similar to previous, and path will not deviate unnecessarily. This will lead the user to belive that he tried to recover and succeeded. This strengthens our primary objective i.e., of seamless transfer of control towards smooth and safe navigation.

Algorithm 1. Margin based Obstacle Avoidance Algorithm

Require: d < D and S=SAFE and n=0 and U$\neq$ 0 and t $\neq$0
Ensure: u (timer has initialized)
1: **while** u $\leq$ U **do**
2:    **if** d < D **then**
3:        S = SAFE
4:    **else**
5:        S=CLOSE
6:        n=n+1
7:    **end if**
8:    **if** u $\leq$ U **then**
9:        **if** n $\geq$ t **then**
10:           S=UnSAFE
11:           u=0
12:           **while** u $\leq$ U **do**
13:               Publish S
14:           **end while**
15:           S=SAFE
16:        **end if**
17:        **if** u$\geq$U **then**
18:           u=0
19:        **end if**
20:    **end if**
21:    Publish S
22: **end while**

'S' stands for state of the Wheelchair, 'd' is width of margin (i.e. safe distance from an obstacle), 'D' is distance of Wheelchair from an obstacle, 'n' is number of mistakes being made in specific time span, 't' indicate threshold (maximum number of mistakes allowed in specific time span), 'u' record the duration of individual mistake, 'U' is the duration for which automated control will drive the wheelchair after frequent mistake within 'u'.

2.2   CCA Within ROS-USARSim

ROS-USARSim is a combined framework for robotic control and simulation [2]. USARSim provides a rich and extensible simulation environment with an efficient physics engine. Pioneer P3AT robot is treated as an intelligent wheelchair and imbibes our control architecture. LCL has been simulated by rossim, it is a part of usarsiminf package distributed within the USARSimROS stack. The 3D indoor environment created in the USARSim simulator is seen as a 2D map using rviz - the ROS visualization tool.

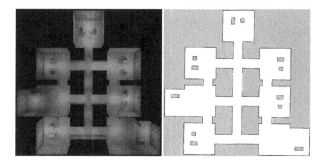

Fig. 4. A sample 3D arena in USARSim and corresponding 2D map in rviz.

A 2D map of the 3D virtual environment withing USARSim has been generated within ROS for visualization using *rviz* (a visualization tool available withing ROS framework shown in Fig. 4) as follows;

```
$ roscore
$ roslaunch usarsim_inf
$ rosrun gmapping slam_gmapping
$ rosrun cca_teleop cca_teleop
$ rosrun map_server map_saver
$ eog ./map.pgm
```

Each from the first command indicates; 1. Executing a hub program of ROS; 2. Executing an interface program between ROS and USARSim to bring up the P3AT robot; 3. Bring up slam_gmapping, which will take in laser scans; 4. Executing program to drive robot (User mode) on ROS; 5. Saved a snapshot map image from SLAM; 6. Running *eog* at same directory which displays *map.pgm* with automatic renewing. Figure 4 shows one of the USARSim 3D arenas on left and corresponding 2D map generate on right.

CCA has several ROS-nodes working concurrently. A launch (*cca.launch*) file bring up all the required node to run the simulation. Figure 5 displays the running nodes created using ROS utility tool *rqt_graph*.

```
$ roslaunch cca_ver2 cca.launch
```

A number of predefined ROS-nodes have been used not only to reduce the complexity of implementation but also to exhaust the basic functionality of the software framework. Which includes the drivers to interact with sensors and actuators, localization through *amcl* etc.

ccaROS introduces nodes for analyzing and publishing the sensor data, robot's state and user's navigational strategy to appropriate ROS topics. Figure 3 is the rqt_graph which shows ROS computation graph of ccaROS. The heuristic node receives the IW's state-related information by subscribing to robo_pos (position coordinates of the wheelchair within the environment), nav_strategy_pub (determines which navigation strategy to follow from Wall Follow, Center Point and Least Angle) and robo_dir (determines in which direction

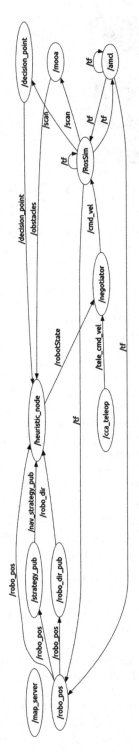

Fig. 5. All the nodes working together for CCA.

the wheelchair is moving) topics and continuously monitors the IW's state. Further data on nav_trategy_pub and robo_dir is being published by strategy_pub and robo_dir_pub node respectively. If IW is found to be unsafe (such as closer to the obstacle) then it calculates the required angle and direction in which IW needs to turn based on the user's navigational strategy. The node sends the angle and direction along with a flag (set to '1' if the robot is in an unsafe state otherwise '0') to the negotiator. If the flag is '1', negotiator blocks the user's velocity commands (sent through cca teleop node) and generates the automated velocity commands; sends them to Local Control Layer. If the flag is '0' negotiator switches drive mode back to cca teleop and let the user control the wheelchair.

3   Experiments and Results

USARSim was administered on an HP z800 workstation computer equipped with Core i3 and 12 GB RAM. The controller application over *ROS-Fuerte* version is administered on another HP z800 workstation of similar configuration and communicates over TCP/IP. Using the ros-node cca_teleop the user drives the robotic wheelchair using arrow keys on a standard 101 computer keyboard.

Finish Time Comparison. 20 participants categorized on two cognitive category: High Cognitive Score (HCS) and Low Cognitive Score (LCS) were grouped [5]. Each were given a trial to drive through the maze in two different mode; with assistance and without assistance. Figure 6 shows that for LCS group participants total finish time is higher, whereas when assistance were provided their drive time are approximately same.

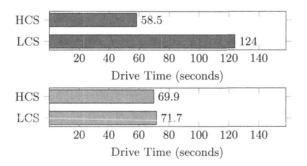

Fig. 6. Total finish time; above without assistance and below with assistance.

Trajectory. Since transfer of control between user and agent should occur in a way that user remains unaware of control transfer, navigation path should not vary unnecessarily. Figure 7 shows the trajectory of the IW; blue line is where the user is driving and red line indicate automode. Path before and after control transfer do not deviate unnecessarily.

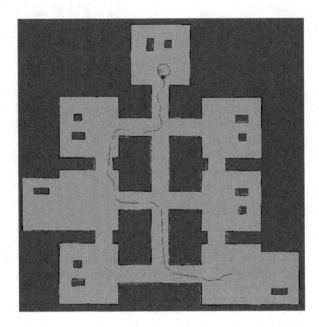

Fig. 7. Trajectory shows a smooth path during negotiation.

Participants Feedback. User gave score for trials in a scale of 1 to 5; where 1 is lowest (i.e. poor) and 5 is highest (i.e. best). Users were not told which trial worked with assistance and which one without. 74% of the users believe that the trial with assistance were highly comfortable. 12% believe that trial with assistance were slightly better and rest did not found any significant difference.

4   Conclusion

This paper has presented a new ROS node as a middle-ware package of CCA. The package provides all the functionality underlined by CCA ranging from both low level and high level navigation, obstacle avoidance as well as mimic users strategy after control transfer. We have examined the functionality of ccaROS within a ROS-USARSim environment and found it satisfactory.

References

1. Anderson, J.R., Bothell, D., Byrne, M.D., Douglass, S., Lebiere, C., Qin, Y.: An integrated theory of the mind. Psychol. Rev. **111**(4), 1036 (2004)
2. Balakirsky, S., Kootbally, Z.: USARSim/ROS: a combined framework for robotic control and simulation. In: ASME/ISCIE 2012 International Symposium on Flexible Automation, pp. 101–108 (2012)
3. Carlson, T., Demiris, Y.: Collaborative control for a robotic wheelchair: evaluation of performance, attention, and workload. IEEE Trans. Syst. Man Cybern. Part B Cybern. **42**(3), 876–888 (2012)

4. Cooper, R.A., Boninger, M.L., et al.: Engineering better wheelchairs to enhance community participation. IEEE Trans. Neural Syst. Rehabil. Eng. **14**(4), 438–455 (2006)
5. Crum, R.M., Anthony, J.C., Bassett, S.S., Folstein, M.F.: Population-based norms for the mini-mental state examination by age and educational level. JAMA **269**(18), 2386–2391 (1993)
6. Langley, P., Choi, D.: A unified cognitive architecture for physical agents. In: Proceedings of the National Conference on Artificial Intelligence, vol. 21, p. 1469. AAAI Press; MIT Press, Menlo Park; Cambridge; London (1999, 2006)
7. Pushp, S., Bhardwaj, B., Hazarika, S.M.: Cognitive decision making for navigation assistance based on intent recognition. In: Ghosh, A., Pal, R., Prasath, R. (eds.) MIKE 2017. LNCS (LNAI), vol. 10682, pp. 81–89. Springer, Cham (2017). https://doi.org/10.1007/978-3-319-71928-3_9
8. Pushp, S., Saikia, A., Khan, A., Hazarika, S.M.: A cognitively enhanced collaborative control architecture for an intelligent wheelchair: formalization, implementation and evaluation. Cogn. Syst. Res. **49**, 114–127 (2018)
9. Quigley, M., et al.: ROS: an open-source robot operating system. In: ICRA Workshop on Open Source Software. vol. 3, p. 5, Kobe (2009)
10. Saikia, A., et al.: cBDI-based collaborative control for a robotic wheelchair. Proced. Comput. Sci. **84**, 127–131 (2016)
11. Schultz, A.: Using computational cognitive models to build better human-robot interaction. In: NAE US FOE Symposium (2006)
12. Urdiales, C., et al.: Wheelchair collaborative control for disabled users navigating indoors. Artif. Intell. Med. **52**(3), 177–191 (2011)

IoT Monitoring of Water Consumption for Irrigation Systems Using SEMMA Methodology

Sandra López-Torres[1], Humberto López-Torres[2], Jimmy Rocha-Rocha[1],
Shariq Aziz Butt[3], Muhammad Imran Tariq[4], Carlos Collazos-Morales[5],
and Gabriel Piñeres-Espitia[1(✉)]

[1] Universidad de la Costa, Barranquilla 080002, Colombia
gpineres1@cuc.edu.co
[2] Instituto Tecnológico de Soledad Atlántico - ITSA, Barranquilla 080002, Colombia
[3] The University of Lahore, Islamabad 54590, Pakistan
[4] Superior University of Lahore, Lahore 53770, Pakistan
[5] Universidad Manuela Beltrán, Bogotá 111321, Colombia

Abstract. The efficient use of water is an issue that has captured the attention of scientists, technicians, and the community at large. The sustainability of water resources has been threatened by the current imbalance between water supply and demand. Intelligent consumption of water would contribute to the balance and reduce the waste in applications such as the agriculture. This paper shows the design of a water consumption monitoring system based on the Internet of Things (IoT). With the implementation of this system could be known in real time the consumption of water in a crop. In addition, the user of the system may take corrective actions that optimize their water consumption; this is achieved by applying the SEMMA methodology to evaluate the data obtained by the system using two cluster algorithms, Simple K-means and GenClus++. With the application of SEMMA it was possible to determine periods of water consumption that were considered as waste in the irrigation of crops, applying data analysis with both algorithms.

Keywords: Internet of Things (IoT) · Irrigation system · Sample Explore Modify Model, and Assess (SEMMA) · Water consumption

1 Introduction

Water consumption is one of the most important global concern. The United Nations (UN) has established in [1], the assurance of availability and sustainable management of water and sanitation for all people; in [1], The UN declared "In 22 countries (mostly in Northern Africa and Western Asia and in Central and Southern Asia), water stress (defined as the ratio of freshwater withdrawn to total renewable freshwater resources) is above 70% (China is above 75%) [2]. This indicates a strong probability of future water scarcity" and "Agriculture accounts for almost 70% of global water withdrawal, which

© Springer Nature Switzerland AG 2020
U. S. Tiwary and S. Chaudhury (Eds.): IHCI 2019, LNCS 11886, pp. 222–234, 2020.
https://doi.org/10.1007/978-3-030-44689-5_20

is projected to increase significantly to meet food needs" [1]; in Colombia, a Latin-American developing country as particular reference, the use combined of agriculture and livestock is above 66% [3]; the above express that the consumption of freshwater on agriculture must be focused to improve the life of the reserves of water. The greater use of water in agriculture could not be other than the irrigation systems. The use of electronics systems to control and study the water consumption of irrigation system, have relevant rise beginning in the 90s decade, we can found notables firsts developments as [4] where describe the implementations to record the length of time water is in contact with the soil surface.

The monitoring work, is the first stage to take decision to improve the performance of irrigation systems in agriculture. The technology use to improve the agriculture, can be called Precision Agriculture [5]. Internet of things has an rising impact use in agriculture, [6–8], and we considering that is a very strong way to attack the waste of freshwater. Thus, this paper shows the design of a system for monitoring water consumption into irrigation system using the Internet of Things (IoT); a data clustering analysis to identify the water consumption waste periods is performed based on the collected data, and a regression that can be used to describe the consumption. With the implementation of this system could be known in real time the water consumption on the irrigation system; this system allows monitoring environmental variables like temperature and humidity, so can be used in the level of application of distribution networks (hydrants), constituting be a part of valid practice for efficient irrigation system, as described in [9]. For the data analysis, Cluster K-Means [10] and GenClus++ [11] algorithms were applied. Three classification tree algorithms were used to analyze the quality of the data grouped in the intervals, showing that GenClus++ describes a real behavior in water consumption.

This article has the following organization: Sect. 2 is a review of related work; Sect. 3 shows the proposed system and methodology for the tests and analyzes the data collected; Sect. 4 shows results and analysis; finally the conclusions are shown.

2 Review of Related Work

2.1 Interests on Improve Water Consumption for Irrigation Systems

The importance of water consumption is originating by the concern for the possible disappearance of fresh water [2]. In [9] is declared that water-efficient practices enhance the economic viability and environmental sustainability of irrigated agriculture, without necessarily reducing water usage, also, enounces technologies, impacts and necessary improvements, useful to classify strategies that wants improve water consumptions, like the proposed in this paper. In [12] is studied the effect of irrigation water on agriculture into a crop in particular, confirming that irrigation water has significant direct effect both crop revenue and receptivity, from this achievement, we considering that the efficient water consumption benefits go beyond that conserving a resource natural.

Other works, are focusing in the management freshwater reserves as it happens in [13], which analyzes the potentiality of the fresh ground water reserve in a littoral, using an integrated approach consisting of geophysical surveys, well measurements, geological and geomorphological characterization, and chemical analyzes. In [14], has been evaluated all the considerations required to implement an irrigation system for sugarcane crops with great impact on a very important water reserve; this paper is relevant because, in some cases the best strategy to improve water consumption is not implement or suppressed irrigation systems. Similarly, in [15] is declared that the fundamental solution to this issue is to develop precision irrigation, according to crop water demand timely information, developing an appropriate irrigation index. In [16] mentions that electronic systems based in Wireless Sensors Networks are important in the handling and management of water resources for irrigation, being useful for understanding the changes in the crops, assess the optimum point for harvesting, estimating fertilizer requirements and to predict crop performance more accurately.

2.2 IoT Applications on Irrigation Systems in Agriculture

In the last decade, several systems aim to optimize the water consumption of irrigation systems in agriculture have been developed. One of these works was developed in Australia, where a sensor-based border verification irrigation system was implemented; the system consists of a wireless sensor, actuation network, central host/user interface (which collects data and displays real-time information) and the central control system software; the system analyzes data and generate reports in real time, plays a double role in the programming and monitoring of irrigation events [17]. In Romania an IoT architecture for irrigation systems in crops is implemented; this system is based in pipes and actuators to allows coverage an area near to 1–5 Ha; this irrigation system consist of a data acquisition platform for monitoring soil variables (temperature, humidity, Ph) and environment variables (temperature, humidity relative, atmospheric pressure, light intensity) and then, the data collected are they send it using Zigbee protocol to a IBM Bluemix IoT platform; trough of agriculture algorithms is possible take decision with the data received [18].

Other work is carried out at Indian, where a smart irrigation system is it built using neuronal network; this system is based on the USP node, which consisting of ESP8266 Module, a soil moisture sensor, and temperature and humidity sensor; the data information is sent to server using the 802.11x protocol; in the server the data is upload to cloud for monitoring and management system; this system is low cost and allows getting accurate for ON/OFF the irrigation system [19]. Also in India, an irrigation system was developed using a WSN consisting of nodes based on the SENSEnuts platform, a soil moisture sensor, a humidity sensor, a temperature sensor and a light intensity sensor; these sensors allow decisions to be made to activate or not activate a motor sprinkler/pump; the WSN uses the Zigbee protocol to send the data to Coordinator node; this node has the Thing Speak tool to upload information in the cloud and can be monitored by the farmer in real time [20]. A new work is proposed in India, which is based irrigation system using Arduino UNO platform, a soil temperature and humidity sensor, a camera, a water flow sensor, a Raspberry Pi and an Xbee S1 module; this system controls a solenoid valve for water irrigation and obtains measurements of water flow, plants health, soil temperature and humidity, which are loaded into the server through the Raspberry Pi for farmer monitoring; the system works in manual or automatic mode [21]. Another work in India proposes a based irrigation system on Arduino UNO platform and an ESP8266 module; this system consist of a LM35 temperature sensor, FC28 moisture sensor and L10530 pH sensor for manages water and fertilizer through motors that drive bombs; the Thinger tool is used as an IoT platform to data monitoring and crops supervisory [22].

At Spain an experimental WSN on the ecological horticultural enterprise is development. The system was successfully implemented on a crop of ecological cabbage (Brassica oleracea); the result was a low cost, highly reliable and simple infrastructure for the collection of agronomical data [16]. In [23], is implemented a WSN, with auto-fuzzy, remote and manual operation to realize real-time monitoring for soil moisture and automatic control of agricultural irrigation.

2.3 Water Consumption Data Mining Analysis

The data analysis is a factor important for the improved of the irrigation systems. At [24] the K-means algorithm is used for classified periods the consumption of water, gas, and electricity; then, tree decision algorithms are used for stablish relation on the characteristic groups found; SVM algorithms is used to classification of consumption per day. In [25] a new technique is developed for the data classification, using domestic water consumption data set; this study is relevant because there are very few works that relate water consumption with data mining. Similar studies like [26, 27] and specially [28], describe patterns and take decision thought of data mining and machine learning appliqued in water consumption. Only [29] correspond a very important use of data mining in agriculture, but nevertheless, the approach to obtain efficiency is related to choosing the best crop for the conditions detected, and not making water use more efficient as such. Attending to the above, using data mining techniques to detect waste of water periods applied specificity on an irrigation, it is a scientific vacuum, barely addressed; in that order of ideas, this paper seeks to help fill that scientific gap, combining IoT monitoring with data mining.

3 Materials and Methods

3.1 Description of Situation

In this work, a sprinkler-based irrigation system prototype connected to a single water supply channel controlled by an automatic irrigator was used; the system has sensors for monitoring the temperature and humidity in the environment, the temperature and humidity inside the enclosures of the irrigator and the water flow for the crop. For the study of irrigation quality, we hypothesized that there are periods of time in which the water supplied can be considered as waste, that is when the water supplied is less than the water required per hectare of crop; according to the operation specifications, 16.5 L are necessary to reach a detectable minimum of increased humidity in the crops. To achieve this measure, it is necessary to maintain a minimum water flow of 1.1 L per minute for a time of 15 min, to consider that the irrigation system operates properly. In this order of ideas, for lower water flow rates, the water supplied can be considered as waste. To optimize its operation, it is proposed to use waste detection periods, so the irrigation system can be turned off during them, improving the efficiency of water consumption without damaging the crops.

3.2 IoT Monitoring of Water Consumption for Irrigation Systems

This work proposes an IoT platform to continuos monitor water consumption. Consisting of a hardware component and a web-based software component, which allows humidity and temperature monitoring in the environment using a DHT11 sensor, and water consumption in periods of 20 min, using a YF-S201 flow sensor. The hardware component consists of sensor control panels, designed specifically for the water flow sensor and for temperature and humidity sensors, based in a Node MCU (ESP8266). Figures 1 and 2 show the architecture of each sensor control panel, and the general description of the IoT platform; The block diagram defines functions, components and system variables. The software component is a web platform that receives the information collected to be displayed in real time.

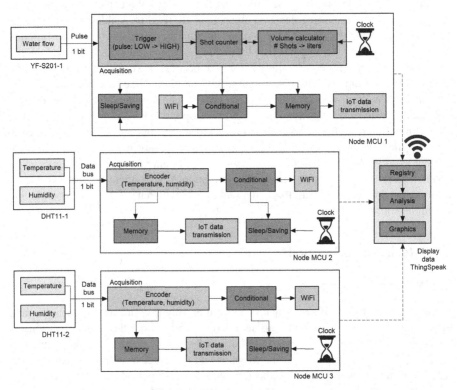

Fig. 1. IoT platform architecture

By the water flow measuring process, the board is capable to detected a unexpected flow activity and the temperature and humidity measuring process, the sensor board send the measures each 5 min; this process is independent of the waterflow measures. If the wireless network connection fail, the boards holds records until the network is available again.

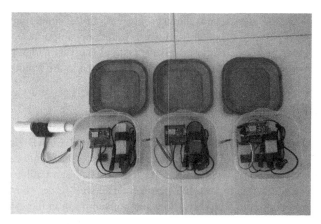

Fig. 2. IoT based irrigation system prototype

The web-based software component is developed using the ThingSpeak IoT service, which receives the data from the sensor boards, and then stores, processes and displays it. To perform the above, ThingSpeak it contains the tools: MATLAB Analysis, MATLAB Visualizations and TimeControl.

3.3   Irrigation Water Consumption Data Analysis

We use a methodology inspired by SEMMA [30], acronym for Sample, Explore, Modify, Model, Asses. This refers to the central process of data mining, starting with a statistically representative sample of the data. SEMMA methodology is applied in this work as follow.

Sample Stage: at this stage of this work, the extract to analyze was the measured consumption data of water for two weeks.

Explore Stage: at this stage of this work, we optimize the comma-separated value files (CSV), the redundant data was deleted and the column data was reorganized, the hours of the day were converted to decimal values between 0 and 1.

Modify Stage: we apply a discretization algorithm of equal width for flow measurements and time of day. Equation 1 gets the width of the intervals w, the Eq. 2 gets the upper interval limits, the Eq. 3 gets the closed intervals to the right, this is done in this way to be able to separate the zero consumption from the near zero consumption. In the case of the hours of the day, the decimal values were discretized in close-left and open-right intervals of 0,083333333 (2 h), for total of 12 intervals, subsequently, they identified the main waste periods, close-left and open-right intervals of 0,010416667 (15 min), will be applied for more precise periods.

$$w = (\text{Flow}_{max} - \text{Flow}_{min})/k \qquad (1)$$

$$w_i = \text{Flow}_{min} + w, \text{Flow}_{min} + 2w, \text{Flow}_{min} + i * w, \cdots, \text{Flow}_{min} + (k-1) * w \qquad (2)$$

$$w_0 = [0, 0], \; w_i = \big(\text{Wmax}_{(i-1)}, \text{Wmax}_i\big] \, 0 < i < k \qquad (3)$$

Model Stage: we use associations and a comparison of the clustering algorithms to find a relationship between time of day and waste water consumption.

Access stage: we performed a descriptive statistical analysis with a box graph and linear regression was applied. The data analysis is disclosed in the results section.

4   Results and Discussions

The platform manages data monitoring information with MATLAB Analytics cloud computing (IoT) tools available on the ThingSpeak platform; monitoring charts and data can be viewed from any device, with any ThingSpeak compatible web browser, resulting in a very accessible remote IoT platform. The web server shows the internal and external temperature and humidity, the water consumption in the last 24 h and the daily water consumption. The user can extract the data in comma-separated values files (.CSV extension file); the humidity and temperature data are extracted separately from water flow data. The system indicate to the minimum and maximum temperature was 21 °C and 33 °C, respectively, with an average of 26 °C. The minimum and maximum humidity was 59% and 94%, with an average of 83%.

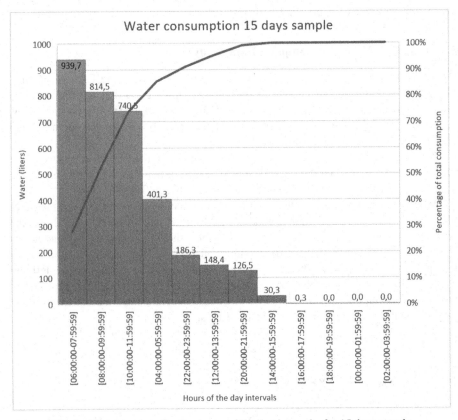

Fig. 3. Water consumption per hours of the day intervals, for 15 days sample

The water consumption data obtained can be used for data mining tools and discover new information. Implementing our SEMMA methodology, during the Model Stage, applying the twelve hours of day intervals, we find that the water consumption is concentrated between 6 and 12 o'clock (6:00:00 to 11:59:59), corresponding to the 73,64% of the total consumption, shown in Fig. 3. Then, new 15-min. intervals were applied from 6 o'clock to 12 o'clock and the water flow intervals shown in Table 1. This Stage found the confidence for seven rules was 100% and 67% for the other rules. The above can be interpreted as that the intervals of water consumption are dis-ordered once they manifest themselves within specific time intervals, so they could be grouped together and, therefore, describe a more general pattern of as-sociation.

Table 1. Water consumption time associated

Cluster	Water consumption associated interval	Time interval associated	Percentage of samples associated
0	(16.5–33]	[07:30:00–07:44:59]	53%
1	(33–49.5]	[10:00:00–10:14:59]	21%
2	(0–16.5]	[08:00:00–08:14:59]	21%
3	(33–49.5]	[06:30:00–06:44:59]	1%
4	(0–16.5]	[07:15:00–07:29:59]	3%

To perform a clustering analysis, K-means clustering analysis was compared with more sophisticated clustering algorithm: GenClus++, which is a clustering algorithm that combining k-means with Genetic Algorithm through a Novel Arrangement of Genetic Operators for High Quality Clustering. This tests were performed with WEKA tool (Table 2).

Table 2. Water consumption-time K-means clustering

Cluster	Water consumption (I)	Main time interval	Samples associated	Class name
0	1,2313	[20:00:00–21:59:59]	18%	Wasting
1	41,9389	[10:00:00–11:59:59]	12%	Normal
2	70,9765	[08:00:00–09:59:59]	11%	High
3	21,7808	[10:00:00–11:59:59]	17%	Accurate
4	10,4405	[08:00:00–09:59:59]	18%	Wasting
5	62,3312	[06:00:00–07:59:59]	10%	High
6	28,0794	[06:00:00–07:59:59]	14%	Accurate

First, for applying k-means we use the next configuration for generate 7 clusters:

```
weka.clusterers.SimpleKMeans -init 0 -max-candidates 100
-periodic-pruning 10000 -min-density 2.0 -t1 -1.25 -t2 -
1.0 -N 6 -A "weka.core.EuclideanDistance -R first-last" -
I 500 -num-slots 1 -S 10
```

For GenClus++ we use the next configuration:

```
weka.clusterers.GenClustPlusPlus -I 60 -P 20 -N 60 -Q 10
-F 50 -D 0.0 -C 50 -S 10
```

The minimum number of clusters generated by GenClus++ was 8 (Table 3):

Table 3. Water consumption-time GenClus++ clustering

Cluster	Water consumption (I)	Main Time interval	Samples associated	Class name
0	2,9064	[08:00:00–09:59:59]	34%	Wasting
1	21,2962	[22:00:00–23:59:59]	8%	Accurate
2	26,3483	[06:00:00–07:59:59]	8%	Accurate
3	42,2938	[10:00:00–11:59:59]	15%	Normal
4	22,0521	[10:00:00–11:59:59]	7%	Accurate
5	61,2026	[06:00:00–07:59:59]	14%	High
6	33,3791	[08:00:00–09:59:59]	10%	Normal
7	23,3945	[20:00:00–21:59:59]	5%	Accurate

K-means detected 2 cluster that we assignats "wasting" class because is under 16,5 L; clusters with same class name can be merged, and the 36% of the total periods of consumption are clustered as wasting. GenClus++ clusters 34% as "wasting", 2% less that simple K-means.

For compare the accuracy of GenClust++ and simple K-means for "wasting" detection periods, were applied the RandomForest, C4.5 (decision tree) and LMT classify algorithms. The Table 4 contains the results of TP rate, precision and recall performance for the wasting class.

After of analyses the box plot and the Table 4, the GenClust++ clustering technique apparently shows a poor performance than K-mens, but this is not true at all. If is analyze the Fig. 4, we found that GenClust++ with the additional analysis of Genetic Algorithm, manages to differentiate that "normal" have a more common appearance that "high" consumption, this is more accurate to the reality. For reduce the wasting periods, during the time of the day detected as wasting, sliding windows on/off routines of the automated irrigator can be programed, to force to reach normal water consumption,

Table 4. Decision tree algorithms performance comparison

Metrics	RandomForest		C4.5 decision tree (J48)		LMT	
	K-means	GenClus++	K-means	GenClus++	K-means	GenClus++
Correctly classified instances	99,145%	91,453%	94,017%	87,1795%	98,290%	95,7265%
Incorrectly classified instances	0,8547%	8,547%	5,9829%	12,8205%	1,7094%	4,2735%
Relative absolute error	7,2935%	15,489%	7,226%	19,3488%	12,939%	17,9696%
TP rate	1,000	1,000	1,000	0,950	1,000	1,000
FP rate	0,000	0,026	0,027	0,052	0,000	0,026
Precision	1,000	0,952	0,955	0,905	1,000	0,952
Recall	1,000	1,000	1,000	0,950	1,000	1,000

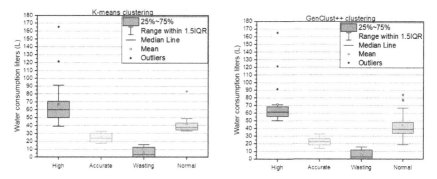

Fig. 4. Box plot diagram for Simple K-Means and GenClus++ clustering

however, complementary studies must be researched to obtain a model to delete waste time of water consumption, based on the detection described in this paper.

Finally, applying a linear regression for the general water consumption, it is obtained a linear behavior described by the function $w(x) = 188.9x + 145.94$, where x is the days passed, and $w(x)$ is the water consumption per day. This linear behavior of the water consumptions allows that any correction or improvement for minimize or eliminate can have a direct impact and can be sustained over time.

5   Conclusions

In this work an irrigation system was design for monitoring the water consumption in a crop, being the most important innovation the detection of waste periods through an analysis SEMMA with the data obtain in the monitoring time. By using SEMMA it is possible stablish times in which the irrigation system is inefficient. A principal contribution in SEMMA was the application of a Simple K-means and GenClust++ clustering algorithm, which allowed to define clusters to detect the periods in which there was waste in the consumption of water for irrigation, it was observe that GenClust++ reach a better description of the reality. This can be very useful to optimize work schedule, to detect malfunction or malicious use of the irrigation system.

The proposal irrigation system is complemented with an IoT based platform which allowed monitoring the water consumption, the humidity and temperature. This is an important function for the farmer for the making decision in the crops.

The irrigation system IoT based is low cost and is easy for their implementation in crops for small farmers, being an important contribution for the development of the agriculture in the region.

References

1. United Nations: The sustainable development goals report 2018. https://doi.org/10.29171/azu_acku_pamphlet_k3240_s878_2016. Accessed 12 Oct 2019
2. Xiong, S.M., Wang, L.M., Qu, X.Q., Zhan, Y.Z.: Application research of WSN in precise agriculture irrigation. In: Proceedings of the 2009 International Conference on Environmental Science and Information Application Technology, ESIAT 2009, vol. 2, pp. 297–300. IEEE, China (2009)
3. Instituto de Hidrología-Meteorología y Estudios Ambientales IDEAM, Ambiente, Reporte de avance del estudio nacional del agua 2018. http://www.andi.com.co/Uploads/Cartilla_ENA_%202018.pdf. Accessed 12 Oct 2019
4. Gohring, T.R., Armstrong, S.D.: Intake opportunity timer for evaluating surface irrigation. Appl. Eng. Agric. **9**, 233–235 (1993)
5. Karim, F., Karim, F., Frihida, A.: Monitoring system using web of things in precision agriculture. Procedia Comput. Sci. **110**, 402–409 (2017)
6. Muangprathub, J., Boonnam, N., Kajornkasirat, S., Lekbangpong, N., Wanichsombat, A., Nillaor, P.: IoT and agriculture data analysis for smart farm. Comput. Electron. Agric. **156**, 467–474 (2019). https://doi.org/10.1016/j.compag.2018.12.011
7. Goap, A., Sharma, D., Shukla, A.K., Rama Krishna, C.: An IoT based smart irrigation management system using machine learning and open source technologies. Comput. Electron. Agric. **155**, 41–49 (2018). https://doi.org/10.1016/j.compag.2018.09.040
8. Caicedo-Ortiz, J.G., et al.: Monitoring system for agronomic variables based in WSN technology on cassava crops. Comput. Electron. Agric. **145**, 275–281 (2018)
9. Levidow, L., Zaccaria, D., Maia, R., Vivas, E., Todorovic, M., Scardigno, A.: Improving water-efficient irrigation: prospects and difficulties of innovative practices. Agric. Water Manag. **146**, 84–94 (2014). https://doi.org/10.1016/j.agwat.2014.07.012
10. Yu, H., Chen, Y., Lingras, P., Wang, G.: A three-way cluster ensemble approach for large-scale data. Int. J. Approximate Reasoning **115**, 32–49 (2019). https://doi.org/10.1016/j.ijar.2019.09.001
11. Islam, M.Z., Estivill-Castro, V., Rahman, M.A., Bossomaier, T.: Combining K-MEANS and a genetic algorithm through a novel arrangement of genetic operators for high quality clustering. Expert Syst. Appl. **91**, 402–417 (2018)
12. Zewdie, M.C., et al.: Direct and indirect effect of irrigation water availability on crop revenue in northwest Ethiopia: a structural equation model. Agric. Water Manag. **220**, 27–35 (2019). https://doi.org/10.1016/j.agwat.2019.04.013

13. Cellone, F., Tosi, L., Carol, E.: Estimating the freshwater-lens reserve in the coastal plain of the middle Río de la Plata Estuary (Argentina). Sci. Total Environ. **630**, 357–366 (2018). https://doi.org/10.1016/j.scitotenv.2018.02.236
14. Perin, V., Sentelhas, P.C., Dias, H.B., Santos, E.A.: Sugarcane irrigation potential in North-western São Paulo, Brazil, by integrating Agrometeorological and GIS tools. Agric. Water Manag. **220**, 50–58 (2019). https://doi.org/10.1016/j.agwat.2019.04.012
15. Li, D., Wang, X.: An intelligent irrigation system for vineyards based on WSN technology: a preliminary study. In: International Conference on Informationization, Automation and Electrification in Agriculture 2008, pp. 733–736 (2008)
16. López Riquelme, J.A., Soto, F., Suardíaz, J., Sánchez, P., Iborra, A., Vera, J.A.: Wireless Sensor Networks for precision horticulture in Southern Spain. Comput. Electron. Agric. **68**, 25–35 (2009). https://doi.org/10.1016/j.compag.2009.04.006
17. Dassanayake, D., Dassanayake, K., Malano, H., Dunn, G.M., Douglas, P., Langford, J.: Water saving through smarter irrigation in Australian dairy farming: use of intelligent irrigation controller and wireless sensor network. In: 18th World IMACS Congress and MODSIM09 International Congress on Modelling and Simulation, pp. 4409–4415 (2009)
18. Dobrescu, R., Merezeanu, D., Mocanu, S.: Context-aware control and monitoring system with IoT and cloud support. Comput. Electron. Agric. **160**, 91–99 (2019)
19. Nawandar, N.K., Satpute, V.R.: IoT based low cost and intelligent module for smart irrigation system. Comput. Electron. Agric. **162**, 979–990 (2019)
20. Joshi, K., Jain, S., Patwari, A.: WSN hardware prototype for irrigation control and multi-parameter plant growth monitoring using IoT. Int. J. Innov. Technol. Explor. Eng. **8**(6), 1643–1650 (2019)
21. Math, A., Ali, L., Pruthviraj, U.: Development of smart drip irrigation system using IoT. In: 2018 IEEE Distributed Computing, VLSI, Electrical Circuits and Robotics, pp. 126–130 (2018)
22. Thakare, S., Bhagat, P.H.: Arduino-based smart irrigation using sensors and ESP8266 WiFi module. In: 2018 Second International Conference on Intelligent Computing and Control Systems (ICICCS), pp. 1–5 (2018)
23. Liu, H.G., Peng, L., You, Y.: ZigBee technology based fuzzy control system for precision irrigation. Adv. Mater. Res. 190–194 (2011). https://doi.org/10.4028/www.scientific.net/AMR.282-283.190
24. Al-Radaideh, Q.A., Al-Zoubi, M.M.: A data mining based model for detection of fraudulent behaviour in water consumption. In: 2018 9th International Conference on Information and Communication Systems, ICICS 2018, pp. 48–54 (2018)
25. Saffie, N.A.M., Rasmani, K.A., Sulaiman, N.H.: Fuzzy classification based on combinative algorithms. In: IOP Conference Series: Materials Science and Engineering, vol. 477, no. 1 (2019)
26. Buttitta, G., Finn, D.: High resolution residential domestic hot water consumption profiles using data mining clustering techniques on time of use data. In: Kaparaju, P., Howlett, R.J., Littlewood, J., Ekanyake, C., Vlacic, L. (eds.) KES-SEB 2018. SIST, vol. 131, pp. 159–168. Springer, Cham (2019). https://doi.org/10.1007/978-3-030-04293-6_16
27. Amroun, H., Benziani, Y., Temkit, M.H., Ammi, M.: Advanced statistical models for modeling hot water consumption using a connected boiler. In: Proceedings - IEEE 2018 International Congress on Cybermatics, pp. 35–42 (2018)
28. Chatzigeorgakidis, G., Karagiorgou, S., Athanasiou, S., Skiadopoulos, S.: FML-kNN: scalable machine learning on Big Data using k-nearest neighbor joins. J. Big Data **5**(1), 1–27 (2018). https://doi.org/10.1186/s40537-018-0115-x
29. Shirsath, R., Khadke, N., More, D., Patil, P., Patil, H.: Agriculture decision support system using data mining. In: Proceedings of the 2017 International Conference on Intelligent Computing and Control, I2C2 2017, pp. 1–5 (2018)

30. Mohd Selamat, S.A., Prakoonwit, S., Sahandi, R., Khan, W., Ramachandran, M.: Big data analytics: a review of data-mining models for small and medium enterprises in the transportation sector. Wiley Interdiscip. Rev. Data Min. Knowl. Discov. **8**, 1–14 (2018). https://doi.org/10.1002/widm.1238

Applications of HCI

Extracting Community Structure in Multi-relational Network via DeepWalk and Consensus Clustering

Deepti Singh[(✉)] and Ankita Verma

Department of CSE and IT, Jaypee Institute of Information Technology, Noida, India
{deepti.singh,ankita.verma}@jiit.ac.in

Abstract. In the real world, entities are often connected via multiple relations, forming multi-relational network. These complex networks need novel models for their representation and sophisticated tools for their analysis. Community detection is one of the primary tools for the structural and functional analysis of the networks at the macroscopic level. Already a lot of research work has been done on discovering communities in the networks with only single relation. However, the research work on discovering communities in multi-relational network (MRN) is still in its early stages. In this article, we have proposed a novel approach to extract the communities in a multi-relational network using DeepWalk network embedding technique and Consensus clustering. Empirical study is conducted on the real-world publicly available Twitter datasets. In our observations we found that our proposed model performs significantly better than some of the baseline approaches based on spectral clustering algorithm, modularity maximization, block clustering and non-negative matrix factorization.

Keywords: Multi-relational network · Community discovery · Network embedding · Social network · Consensus clustering

1 Introduction

In a network when similar sort of nodes linked with one another through more than one relation then Multi-relational network (MRN) is formed. These kinds of networks are very dominant in real-world scenarios such as social networks, information networks, economic activity network, and bibliographic networks, etc. [1]. In such networks, a node can be a user, an author, a group or any other entity depending upon the kind of network. The relations among the nodes can be of friendship, kinship, authorship, collaboration etc. Figure 1 illustrates the examples of multi-relational network. A set of users from a website YouTube, can be connected via different types of relations as shown in Fig. 1(a). In Fig. 1(b), the multi-relational network formed by integrating information about the same set of users from different websites is illustrated. Due to multiple relations, the network complexity increases tremendously. Various groups of nodes can be formed to make a community considering all the relations among them at a time. Community discovery in networks finds a significant place in simplifying the complexities of such networks [2].

© Springer Nature Switzerland AG 2020
U. S. Tiwary and S. Chaudhury (Eds.): IHCI 2019, LNCS 11886, pp. 237–247, 2020.
https://doi.org/10.1007/978-3-030-44689-5_21

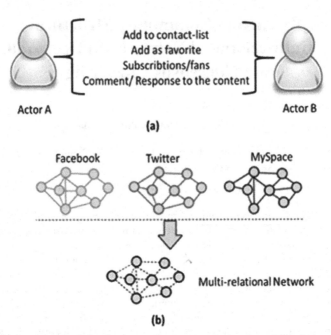

Fig. 1. Examples of multi-relational network. (a) Multiple types of possible relationships between the users on YouTube website (b) A multi-relational network formed by the content of multiple websites.

Single relational network (SRN) can only provide limited information about the nodes. However, MRN provides more information about the nodes and hence communities formed in this are much meaningful and practical. To meet this end, we have proposed a novel framework to extract communities aka cohesive sub-groups in a given MRN. We have adopted a two-fold approach. In the first step, we have learned the latent feature of network nodes from every relation of the MRN using DeepWalk [3] network embedding technique. After learning the node features, the second step applies k-means clustering to find out clustering with respect to each relation. Subsequently, we have integrated all the resultant relation specific clustering results using a consensus clustering framework to reveal the community structure encompassing all the relations of the network [4].

The content of this article is organized as follows: Sect.2 presents the related work. In Sect. 3, we have described the proposed framework. Section 4 discusses the details of the experimental setup and the obtained results. Finally, major findings and future directions are presented in Sect. 5.

2   Related Work

Community discovery algorithms based on inference model [5], modularity [6] of the network and spectral analysis [7] for SRNs have been discussed from long time, but for MRNs there is still a lot to research and analyze. Liu et al. [8] optimizes the composite modularity in a network for discovering the underlying community structure. However, this approach is not suited for large-scale networks. A community extraction method MetaFac (MetaGraph Factorization) employing tensor factorization is proposed in [9]. Another work based on tensor factorization was suggested in [10, 11]. A novel algorithm Query Exploration is proposed in [12]. This algorithm explores the generated Galois Lattices for extracting academic sub-groups in MRN.

Hafiz et al. [13] used a joint variational Bayes method in discovering different types of communities in multi-relational network. Some researchers have suggested to extract communities individually in every layer of the network and then use consensus clustering approach to get the final clustering result [14, 15]. Other methodology includes enhancing single-relation networks based on the generative model like -Stochastic Block Model approach for multi-relational networks [16–19]. Few authors have proposed a modularity index, which characterizes the communities in multilayer networks [20]. The model proposed in [20] have tested their approach by replacing vanilla modularity with their proposed modularity index in traditional community discovery methods. In [21], the authors have suggested an idea for multi-layer graph clustering with edge labels (MiMAG). This idea was later extended in [22] to extract shared cliques in multilayer graphs.

3 Proposed Framework for Community Detection

This section presents the problem statement with detailed description of the proposed scheme.

3.1 Problem Definition

In an MRN, a set of similar types of entities are connected via multiple types of relations. Different relations are modeled by using different graphs. An MRN can be formally defined as

$$G_{MNET} = \{G_i | 1 \leq i \leq r, G_i = (V, E_i)\} \tag{1}$$

where G_i is the graph of i-th relation network in which a vertex set V is linked through the edge set E_i, r is number of relations in the given MRN. The vertex set V comprises of all entities in the network. Let the number of entities be $n = |V|$. Given the number of communities k, the task of community discovery is to obtain a community structure $C \in \{0, 1\}^{N \times k}$ shared across all the r relations in the network. Each entry $C(i, j) = 1$ indicates that i-th entity in the network falls in j-th community. Since, we are interested in finding disjoint community structure, an entity may belong to at-most one community at a time.

3.2 Proposed Approach

Our proposed approach comprises two steps as mentioned below:
Step 1: Learning latent representation of the entities in network.

In this step, we apply DeepWalk [3] algorithm on each relation of MRN to learn latent feature representation of the vertices. In our proposed approach, we have employed DeepWalk on each Graph G_i to generate the network embedding $F_i \in \mathbb{R}^{N \times d}$, where d is the dimension of the representation. The obtained feature matrix F_i can be used to develop various statistical models for in-depth analysis of the graph structure. The outcome of this step is the latent feature matrices F_i for each $1 \leq i \leq r$.

DeepWalk is a deep learning-based approach that learns the latent social features from the graph by performing truncated random walks in the graph. It extends the notion of language models to the network embedding, treating the random walks as the sentences.

Step 2: Apply Consensus clustering approach to extract the community structure.

Once the latent features corresponding to each relation in MRN are revealed, we attempt to unveil shared consensus community structure. On each of the feature matrix F_i corresponding to each relation in MRN, we first apply k-means algorithm to obtain the individual clustering result C_i. In order to mitigate, the seed sensitivity of the k-means algorithm, the initial seeds are selected by k-means++ algorithm [23]. This algorithm first chooses a random data-point as the first initial center and then selects each subsequent center, according to the probability proportional to its distance to the already chosen closest cluster center. Then, the obtained ensemble of clustering results is aggregated into a final partition using the further explained consensus function.

From each of the clustering result C_i, we construct a co-occurrence matrix M_i of size $N \times N$ such that

$$M_i(x, y) = \begin{cases} 1 & \text{if } x^{th} \text{ and } y^{th} \text{ data} - \text{points} \\ & \text{are in the same cluster} \\ 0 & \text{otherwise} \end{cases} \tag{2}$$

The overall consensus matrix which represents the proportion in which the two data-points, co-occur in a cluster is given as:

$$M = \frac{1}{r} \sum_{i=1}^{r} M_i \tag{3}$$

To obtain the final community structure $C \in \{0, 1\}^{N \times k}$, we have applied hierarchical agglomerative clustering with mean linkage on consensus matrix M [24]. In Hierarchical Agglomerative Clustering (HAC), a bottom up approach is adopted to construct nested partitions of the data-points such that a cluster contains other sub-clusters. To begin with, singleton clusters are constructed with each data-point. At each level of hierarchy, distances between all the pairs of clusters are computed. In mean linkage HAC, the mean distance between every point in one cluster to every point in the other cluster is taken as the distance between two clusters [25]. Further, two closest clusters are combined together to form a cluster up in the hierarchy until desired number of communities are formed. Figure 2 illustrates the algorithm of the proposed scheme.

Input:

Multi-relational network $G_{MNET} = \{G_i \mid 1 \leq i \leq r\}$

Number of entities N

Number of relations R

Number of communities k

Number of latent dimensions d

Walk length t

Window size w

Walk per vertex γ

Output:

Community structure $C \in \{0,1\}^{N \times k}$

Algorithm:

1. For i=1 to r do
2. $F_i = Deepwalk(G_i, w, \gamma, d, t)$;
3. end
4. For i=1 to r do
5. $C_i = kmeans(F_i, k)$
6. end
7. Initialize a null matrix M of size $N \times N$
8. For i=1 to r do
9.   For each pair of x and y data-points
10.   that co-occur in the clustering result C_i
11. $M(x,y) = M(x,y) + \frac{1}{r}$
12.   end
13. end
14. $C = HeirAggloClustering(M,k)$
15. Return C

Fig. 2. Algorithm for the proposed approach of community detection.

4 Empirical Evaluation

This section presents the experimental study and the obtained results.

4.1 Dataset Description and Evaluation Metrics

We have performed the experimental study on the real-world Twitter datasets. Twitter dataset is a collection of five datasets viz Football, PoliticsUK, PoliticsIE, Olympics and Rugby [26]. Entities in each dataset are connected via three types of relations: "follows", "mentions", and "retweets". All the relations are directed and binary. The statistical description of the datasets is given in Table 1. As the number of desired communities is given a priori with the data, the following external evaluation metrics are used for evaluating the performance of the proposed approach.

Let $CS' = \left\{C_1', C_2', \ldots, C_K'\right\}$ be the actual community structure and $CS = \{C_1, C_2 \ldots, C_K\}$ be the predicted community structure. The adopted evaluation metrics are described as below:

Purity: This metric measures the proportion of data-points belonging to one class are clustered in one community. The value of purity ranges between 0-1. The larger the value, better the clustering solution is.

Entropy: This measures the extent of randomness in the data. The larger value indicates more randomness, hence poor cluster formation.

Table 1. Summary Statistics of Datasets

Datasets	#Users	#Communities	Density		
			Follows	Mentions	Retweets
Football	268	20	8.60×10^{-2}	7.40×10^{-2}	3.45×10^{-2}
PoliticsIE	348	7	2.07×10^{-1}	7.98×10^{-2}	4.07×10^{-2}
PoliticsUK	419	5	2.27×10^{-1}	1.29×10^{-1}	6.92×10^{-2}
Olympics	464	28	7.23×10^{-2}	6.21×10^{-2}	2.64×10^{-2}
Rugby	854	15	6.27×10^{-2}	5.94×10^{-2}	2.64×10^{-2}

Normalized Mutual Information (NMI): The NMI between CS and CS' is given as:

$$NMI(CS, CS') = \frac{\sum_{i=1}^{K} \sum_{j=1}^{K} n_i^j \log\left(\frac{n.n_i^j}{n_i.n_j}\right)}{\sqrt{\left(\sum_{i=1}^{K} n_i \log \frac{n_i}{n}\right)\left(\sum_{j=1}^{K} n_j \log \frac{n_j}{n}\right)}} \tag{4}$$

where, total number of entities $= n$, $|C_i'| = n_i$, $|C_j| = n_j$ and $n_i^j = |C_i' \cap C_j|$. Higher the NMI, closer will be the partition to the ground truth.

F-score: Let T denotes the set of those pairs of data-sample that are in the same class in CS. Similarly, S represents the set such pairs that belong to the same cluster in CS'. The F_score can be computed as follows:

$$precision = \frac{|S \cap T|}{|S|} \quad recall = \frac{|S \cap T|}{|T|} \tag{5}$$

$$F\_score = \frac{2 \times precision \times recall}{precision + recall} \tag{6}$$

Higher value of these evaluation metrics indicates better performance.

4.2 Experimental Setup and Results

For a comparative analysis of the proposed scheme, we have compared its performance against the following baseline algorithms:

- **SpecCPSA**: In this method, Spectral clustering algorithm [27] is used to obtain individual clustering results that are then combined using Cluster-Based Similarity Partitioning Algorithm (CPSA) [28].
- **SpecLink**: Similar to the previous method, relation-specific clustering results are obtained by spectral clustering, but consensus function used is the mean linkage on co-association matrix as data [24].
- **ModCPSA**: Similar to the SpecCPSA but clustering results are obtained by Modularity maximization [6].

Table 2. Performance of the methods on various datasets

Datasets	Evaluation Metrics	Methods SpecCPSA	SpecLinkage	BlockCPSA	BlockLinkage	ModCPSA	ModLinkage	NNMFCPSA	NNMFLinkage	DW-CPSA	DW-Linkage
Football	Entropy	2.774	2.709	2.377	2.125	2.048	2.157	2.462	2.300	0.697	**0.667**
	Purity	0.310	0.326	0.429	0.479	0.488	0.470	0.398	0.438	0.803	**0.824**
	NMI	0.373	0.385	0.496	0.536	0.555	0.530	0.481	0.505	0.844	**0.848**
Olympics	Entropy	3.118	3.112	2.294	2.142	2.152	2.105	2.227	2.043	0.535	**0.496**
	Purity	0.294	0.306	0.472	0.505	0.512	0.519	0.489	0.531	0.848	**0.872**
	NMI	0.361	0.371	0.536	0.558	0.560	0.566	0.551	0.579	0.877	**0.882**
PoliticsIE	Entropy	2.205	2.204	1.327	1.415	1.416	1.472	1.561	1.567	0.491	**0.426**
	Purity	0.423	0.424	0.628	0.607	0.640	0.598	0.594	0.580	0.868	**0.889**
	NMI	0.058	0.059	0.401	0.363	0.369	0.345	0.322	0.315	0.740	**0.753**
PoliticsUK	Entropy	1.566	1.566	1.188	1.178	1.071	1.107	1.126	1.043	**0.229**	0.307
	Purity	0.473	0.473	0.657	0.650	0.687	0.674	0.688	0.718	**0.948**	0.935
	NMI	0.036	0.036	0.234	0.236	0.300	0.276	0.275	0.316	**0.778**	0.723
Rugby	Entropy	2.183	2.150	2.048	1.881	1.917	1.904	1.996	1.977	1.051	**1.035**
	Purity	0.466	0.480	0.513	0.566	0.560	0.560	0.533	0.542	0.795	**0.800**
	NMI	0.382	0.413	0.367	0.406	0.398	0.395	0.391	0.392	0.630	**0.632**

Bold faces denote the best performance

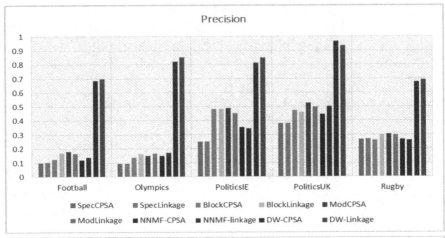

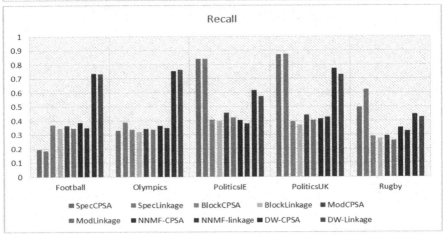

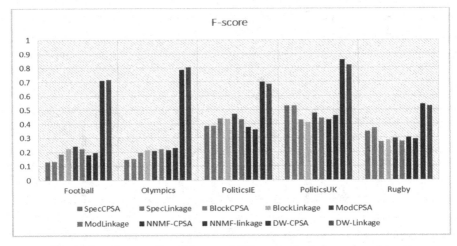

Fig. 3. Precision, recall and F-score of different methods for various datasets

- **ModLinkage**: Akin to previous method, but clusters are combined using mean linkage on the co-association matrix as data.
- **BlockCPSA**: This method combines block clustering with the CPSA.
- **BlockLinkage**: This method is the combination of block clustering with mean linkage clustering.
- **NNMF-CPSA**: Non-negative matrix factorization [29] is applied to reveal latent features. Consensus mechanism used is CPSA.
- **NNMF-Linkage:** Akin to previous method, but clusters are combined using mean linkage on the co-association matrix as data.
- **DW-CPSA:** This method is similar to our proposed approach, but consensus mechanism used here is CPSA.

In our proposed approach DW-Linkage, the different parameters for DeepWalk algorithm are stated as follows: number of latent dimensions $d = 64$, walk length $t = 80$, window size $w = 10$, walk per vertex $\gamma = 40$. The number of communities k for each dataset is set equal to the number of ground truth communities. To avoid any random error that may creep in, all the algorithms are run 10 times and the average values are reported.

Table 2 shows the values of purity, entropy, NMI for all the competing methods. As observed from the Table 2, DeepWalk based algorithms DW-CPSA and DW-Linkage outperform all the other methods. Thus, re-instating the fact that DeepWalk algorithm learns robust features from the network. It is also observed that using hierarchical clustering with mean linkage on the co-association matrix as the consensus function achieves better performance than CPSA algorithm. The results obtained are consistent with the findings of reference. Figure 3. shows the precision, recall and F-score values of different methods. It can be observed from the figure that almost in all the cases, our proposed algorithm outperforms the other algorithms.

5   Conclusion and Future Direction

This article proposes a DeepWalk based approach for community discovery in MRNs. Our proposed approach is a two-fold mechanism that first learns latent feature representation corresponding to every relation in the network. These revealed feature matrices are then used to obtain different clustering results, which are subsequently integrated employing a consensus clustering approach. The proposed method outperforms various baseline techniques in the experimental study conducted on publicly available twitter datasets.

In the proposed algorithm, the number of communities needs to be specified beforehand. However, in real-world applications, the optimal number of communities may not be known a-prior. In the future, we will attempt to determine the number of clusters automatically. We will also consider extending the work for detection of overlapping communities. Further, it will be interesting to see how other state-of-the-art network embedding techniques can be applied in lieu of DeepWalk.

References

1. Cai, D., Shao, Z., He, X., Yan, X., Han, J.: Community mining from multi-relational networks. In: Jorge, A.M., Torgo, L., Brazdil, P., Camacho, R., Gama, J. (eds.) PKDD 2005. LNCS (LNAI), vol. 3721, pp. 445–452. Springer, Heidelberg (2005). https://doi.org/10.1007/11564126_44
2. Kim, J., Lee, J.G.: Community detection in multi-layer graphs: a survey. ACM SIGMOD Rec. **44**(3), 37–48 (2015)
3. Perozzi, B., Al-Rfou, R., Skiena, S.: Deepwalk: online learning of social representations. In: Proceedings of the 20th ACM SIGKDD International Conference on Knowledge Discovery and Data Mining, pp. 701–710. ACM (2014)
4. Vega-Pons, S., Ruiz-Shulcloper, J.: A survey of clustering ensemble algorithms. Int. J. Pattern Recogn. Artif. Intell. **25**(03), 337–372 (2011)
5. Hastings, M.B.: Community detection as an inference problem. Phys. Rev. E **74**(3), 035102 (2006)
6. Newman, M.E.: Modularity and community structure in networks. Proc. Nat. Acad. Sci. U.S.A. **103**(23), 8577–8582 (2006)
7. Newman, M.E.: Spectral methods for community detection and graph partitioning. Phys. Rev. E **88**(4), 042822 (2013)
8. Liu, X., Liu, W., Murata, T., Wakita, K.: A framework for community detection in heterogeneous multi-relational networks. Adv. Complex Syst. **17**(06), 1450018 (2014)
9. Lin, Y.R., Sun, J., Castro, P., Konuru, R., Sundaram, H., Kelliher, A.: Metafac: community discovery via relational hypergraph factorization. In: Proceedings of the 15th ACM SIGKDD international conference on Knowledge discovery and data mining, pp. 527–536. ACM (2009)
10. Verma, A., Bharadwaj, K.K.: Identifying community structure in a multi-relational network employing non-negative tensor factorization and GA k-means clustering. Wiley Interdisc. Rev. Data Min. Knowl. Discovery **7**(1), e1196 (2017)
11. Verma, A., Bharadwaj, K.K.: Discovering communities in heterogeneous social networks based on non-negative tensor factorization and cluster ensemble approach. In: Prasath, R., Vuppala, A.K., Kathirvalavakumar, T. (eds.) MIKE 2015. LNCS (LNAI), vol. 9468, pp. 150–160. Springer, Cham (2015). https://doi.org/10.1007/978-3-319-26832-3_15
12. Guesmi, S., Trabelsi, C., Latiri, C.: CoMRing: a framework for community detection based on multi-relational querying exploration. Procedia Comput. Sci. **96**, 627–636 (2016)
13. Ali, H.T., Liu, S., Yilmaz, Y., Couillet, R., Rajapakse, I., Hero, A.: Latent heterogeneous multilayer community detection. In: ICASSP 2019–2019 IEEE International Conference on Acoustics, Speech and Signal Processing (ICASSP), pp. 8142-8146. IEEE, April 2019
14. Burgess, M., Adar, E., Cafarella, M.: Link-prediction enhanced consensus clustering for complex networks. PLoS ONE **11**(5), e0153384 (2016)
15. Lancichinetti, A., Fortunato, S.: Consensus clustering in complex networks. Sci. rep. **2**, 336 (2012)
16. Holland, P.W., Laskey, K.B., Leinhardt, S.: Stochastic blockmodels: first steps. Soc. Netw. **5**(2), 109–137 (1983)
17. Valles-Catala, T., Massucci, F.A., Guimera, R., Sales-Pardo, M.: Multilayer stochastic block models reveal the multilayer structure of complex networks. Phys. Rev. X **6**(1), 011036 (2016)
18. Reyes, P., Rodriguez, A.: Stochastic blockmodels for exchangeable collections of networks. arXiv preprint arXiv:1606.05277 (2016)
19. Barbillon, P., Donnet, S., Lazega, E., Bar-Hen, A.: Stochastic block models for multiplex networks: an application to a multilevel network of researchers. J. Roy. Stat. Soc. Ser. A (Stat. Soc.) **180**(1), 295–314 (2017)

20. Pramanik, S., Tackx, R., Navelkar, A., Guillaume, J.L., Mitra, B.: Discovering community structure in multilayer networks. In: 2017 IEEE International Conference on Data Science and Advanced Analytics (DSAA), pp. 611–620. IEEE, October 2017
21. Boden, B., Günnemann, S., Hoffmann, H., Seidl, T.: Mining coherent subgraphs in multi-layer graphs with edge labels. In: Proceedings of the 18th ACM SIGKDD International Conference on Knowledge Discovery and Data Mining, pp. 1258–1266. ACM (2012)
22. Zeng, Z., Wang, J., Zhou, L., Karypis, G.: Coherent closed quasi-clique discovery from large dense graph databases. In: Proceedings of the 12th ACM SIGKDD International Conference on Knowledge Discovery and Data Mining, pp. 797–802. ACM (2006)
23. Arthur, D., Vassilvitskii, S.: k-means ++: the advantages of careful seeding. In: Proceedings of the Eighteenth Annual ACM-SIAM Symposium on Discrete Algorithms, pp. 1027–1035. Society for Industrial and Applied Mathematics (2007)
24. Kuncheva, L.I., Hadjitodorov, S.T., Todorova, L.P.: Experimental comparison of cluster ensemble methods. In: 2006 9th International Conference on Information Fusion, pp. 1–7. IEEE (2006)
25. Rokach, L., Maimon, O.: Clustering methods. In: Maimon, O., Rokach, L. (eds.) Data Mining and Knowledge Discovery Handbook, pp. 321–352. Springer, Boston (2005). https://doi.org/10.1007/b107408
26. Greene, D., Cunningham, P.: Producing a unified graph representation from multiple social network views. In: Proceedings of the 5th Annual ACM Web Science Conference, pp. 118–121. ACM (2013)
27. Von Luxburg, U.: A tutorial on spectral clustering. Stat. Comput. **17**(4), 395–416 (2007)
28. Strehl, A., Ghosh, J.: Cluster ensembles—a knowledge reuse framework for combining multiple partitions. J. Mach. Learn. Res. **3**, 583–617 (2003)
29. Lee, D.D., Seung, H.S.: Algorithms for non-negative matrix factorization. Adv. Neural. Inf. Process. Syst. **13**, 556–562 (2001)

Virtual-Reality Training Under Varying Degrees of Task Difficulty in a Complex Search-and-Shoot Scenario

Akash K. Rao[1]([⊠]), Jibraan Singh Chahal[2], Sushil Chandra[3], and Varun Dutt[1]

[1] Applied Cognitive Science Laboratory, Indian Institute of Technology Mandi, Mandi, India
akashrao@live.com, varun@iitmandi.ac.in
[2] GWC Whiting School of Engineering, Johns Hopkins University, Baltimore, USA
jibraansinghchahal@gmail.com
[3] Institute of Nuclear Medicine and Allied Sciences, DRDO, Delhi, India
sushil.inmas@gmail.com

Abstract. The type of training in virtual-reality (VR) environment plays a crucial role in enhancing military personnel's decision-making ability. Little is currently known about how exposure to different types of training in VR designs may assist operators in getting trained in different simulated scenarios. We developed a VR search-and-shoot simulation with two scenarios in task complexity (novice and professional). Thirty healthy subjects played both the novice and professional scenarios in the VR design. Half of the participants were given novice training first, and half of the participants were given professional training first. We took various cognitive and behavioral measures into consideration for statistical analyses. Results disclosed that the participants who faced the professional scenario first fared better than the participants who faced the novice scenario first. We discuss the implication of our results involving VR technologies for creating effective environments for training military personnel.

Keywords: Virtual reality · NASA-TLX · Novice · Professionals · Training · Behavior · Cognition

1 Introduction

Ensuring cognitive readiness is essential for efficient performance in modern day conflicts. One important aspect of cognitive readiness is related to the development, evaluation, and improvement of virtual reality (VR) displays as a means of training personnel on specific tasks/missions [1]. VR is the use of computer graphics to enable an individual to interact with a synthetic three-dimensional environment [2]. VR technology has been used to advance many fields, including medicine, education, design, training, and entertainment [3]. VR allows the possibility for the individual to immerse himself/herself into the virtual world which allows him/her to better build a mental model of the immediate environment and by seamlessly moving around the virtual scene and examining its descriptors from all possible viewpoints [4]. The reason for VR's effectiveness is

U. S. Tiwary and S. Chaudhury (Eds.): IHCI 2019, LNCS 11886, pp. 248–258, 2020.
https://doi.org/10.1007/978-3-030-44689-5_22

because the human brain manifests the virtual world as close to the real world and this manifestation enables the transfer of the acquired skills to the real world [4, 5]. Virtual environments have the capability of enhancing both internally and externally valid experimental designs [5]. They have found to be valuable in the experimental investigation of behavioral processes (e.g., emotion, attention, behavioral disinhibition) underlying psychiatric disorders [6]. Although the VR technologies are developed to support operations in tactical situations, an evaluation of these technologies and their interaction with human-perceptual cognitive functions and ergonomic constraints is lacking and much needed in literature [7].

Prior research has suggested that difficult objectives during training can serve as a stimulus for strategy development [9]. Hence, employing a complex task and difficult objectives during training might lead to improved performance at transfer because of both cognitive and motivational processes [10]. Adding variable difficulty/complexity during training could have different consequences. The difficulty created could siphon cognitive resources from the primary objective [10]. On the contrary, the difficulty created could force the individuals to engage in more effortful, intricate or extensive processing for fulfilling the objective in hand [11]. Under this assumption, adding a variable difficulty to a task might enhance the learning [12]. According to the training difficulty hypothesis [12], when training is conducted under difficult conditions, the transfer of skills acquired during training is optimal [12]. The differences in the processing demands of tasks is a ramification of the task structure, while differences in the resources that the individuals brings to the tasks can be attributed to individual differences [9]. Task complexity according to [9], is the confluence of various cognitive processes like attention, memory, reasoning among other information processing demands imposed by the structure of the task on the individual [13].

Instance-based learning theory (IBLT) [14], theory of how individuals make decisions from experience, has elucidated decision-making in complex tasks very well. According to IBLT, decision-making is a five-step process: recognition of the situation, the judgment based in experience, choices among options based upon judgments, execution of chosen actions (decision-making), and feedback to those experiences that led to the chosen actions [14]. As per IBLT, and the "difficulty hypothesis" by [10], when the difficulty level of the task is higher, the decision-maker would be able to collect and store more "instances" from the training task, and would get a better mental representation of the objective to be achieved and the series of actions to be executed to efficiently achieve the objective. Thus, based upon IBLT, difficult training would subsequently enhance decision-making [14].

In what follows, first, we provide a brief overview of the research involving VR and their application in the recent past. Next, we detail VR design that differed in task difficulty (novice and professional). Next, we use these display designs in an experiment for evaluating our expectations related to training and transfer. Then, we discuss the results obtained from the experiment and finally highlight the implications of our results for creating effective environments for training military personnel.

2 Background

There have been certain studies conducted that have evaluated human perceptual-cognitive functions and ergonomic constraints against VR display designs in dynamic tasks [15]. For example, [15] described an experiment to evaluate two different types of VR interfaces; the immersive VR head mounted display (HMD) interface with interactive fidelity restricted and the non-immersive high-resolution desktop VR interface in a spatial learning task. Results showcased that higher visual fidelity of the immersive HMD VR interface did not lead to a significantly better performance in the spatial learning task. The participants also revealed higher mental workload requirement and higher oculomotor symptoms in the immersive HMD VR interface compared to the non-immersive high-resolution desktop VR interface [15].

Reference [16] elucidated the effects of immersion, resolution and visual fidelity on object location learning and recall. Like [15], the participants were randomly assigned to execute the task in either an immersive HMD VR interface or a non-immersive high-resolution desktop VR interface [16]. Participants were asked to place three objects in a photorealistic virtual environment of a physical laboratory and later instructed to locate these objects in the real world. In contrary with [15], learning in HMD VR interface resulted in significantly better performance compared to the desktop VR interface.

Researchers in [8] evaluated the influence of indirect vision (IVD) and immersive VR training under varying manned/unmanned interfaces in a complex search-and-shoot simulation. Results revealed that the participants performed better in VR compared to IVD. Hence, VR could be used as a viable and reliable platform for assessing and evaluating the personnel's cognitive performance by imbibing them in near real-life like situations [8].

Researchers in [17] evaluated the efficacy of the training difficulty hypothesis [12] on short-term memory encoding and retrieval across two different sets of experiments. The interval between successive stimuli and the number of memory items to be retrieved were manipulated to vary the task difficulty. Results revealed that as the task difficulty during practice increased, the final test performance increased.

3 Materials and Methods

We describe an experiment to determine the performance at transfer in different VR designs due to varying degrees of task difficulty during training.

3.1 Participants

A total of 30 individuals (16 males and 14 females; mean age = 20.9 years, SD = 1.21 years) at the Indian Institute of Technology (IIT) Mandi, Himachal Pradesh, India took part in this study. The study was approved by the institute ethical committee at IIT Mandi. Participation in the task was entirely voluntary. The participants gave a written consent form before they executed the task. All the participants were from Science, Technology, Engineering, or Mathematics backgrounds. The participants received a flat payment of INR 40 for their participation in the study. Also, the participants could earn a performance-based incentive up to INR 20 per simulation. The performance-based incentive was given to the participants if they successfully executed all the objectives in the simulation.

3.2 The Search-and-Shoot Simulation

A terrain-based virtual search-and-shoot simulation was designed using Unity 3D version 5.4.2 [18], and the 3D avatars of the enemies and the gun were designed using Blender Animation [19]. Figure 1(a) shows an overhead map of the terrain. The participants were subjected to a VR simulation with three headquarters located at different locations in the environment, as shown in Fig. 1(a). The participant's objective was to kill all the enemies and reacquire all the three headquarters which the enemies had infiltrated within a specified time limit (10 min). The participant's health was initialized to 100 and this would decrease based on the complexity of the scenario. The participant's health could regenerate to some extent if he/she do not take in any damage for some time. The participant possessed a gun with variable number of bullets (explained in Table 1). The total number of enemies in the simulation was kept to 16. All enemies' health was initialized to 100 and an enemy was considered weak if the health fell below 50 points. At this point, a random number between 0 to 100 would be automatically generated and checked against the existing value. If the value was greater, then the enemy would keep on fighting and if the value was lesser, the enemy would go into the retreat state. The VR simulation was executed using an Android-based system, using a 5.5-in. Xiaomi Redmi Note 3 smartphone and MyVR HMD goggles. The participants used a DOMO MagicKey Bluetooth Controller to navigate and shoot in the VR simulation.

3.3 Procedure

After the participants filled out their consent form and the demographic questionnaire, they were subjected to an acclimatization phase of five minutes. In the acclimatization phase, participants became familiar with the controls, VR task, and the task objectives.

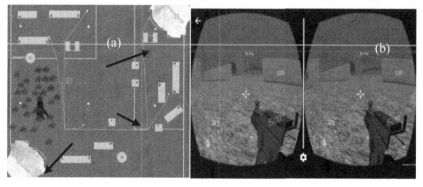

Fig. 1. (a) The overhead map of the terrain designed in Unity3D. The arrows indicate the three headquarters in the simulation. (b) The immersive VR display design.

3.4 Experimental Design

The VR simulation was an immersive 3D virtual reality, catering to a full spectrum view of the artificial environment (see Fig. 1(b)). The simulation provided two training scenarios of varying difficulty in the enemy's artificial intelligence (novice and professional) using state machines and probabilistic networks of variable complexities. Due to the benefits of challenging training over more natural training, we expected improved performance after professional training compared to novice training. In a lab-based setting, all the 30 healthy subjects played both the novice and the professional scenarios across a VR display design (the novice and professional scenarios were presented in a within-subjects format). Half of the participants were given the novice training first (and professional training second), and half of the participants were given professional training first (and novice training second). Various cognitive measures like the computerized version of the NASA-TLX [20] and behavioral variables like number of enemies killed, accuracy index (calculated by dividing the number of bullets taken to kill the enemy by the number of bullets needed to kill the enemy), health index, total time taken to complete the simulation were recorded during the task.

3.5 The Variation in Task Complexity

The variation in the physical attributes of the simulation with respect to the complexities is as shown in Table 1. As shown in Table 1, we introduced some variations in the physical parameters of the task for the variation in task difficulty. The rate of health decrease of the enemy per successful shot fired by the subject was kept to 10 in the novice condition whereas it was kept to four in the professional condition. Similarly, the rate of health decrease of the layer per shot was kept to 2 in the professional condition and it was kept to 1 in the novice condition. We used finite state machines to determine the actions of the enemies. Each major strategy (assault, stealth) would be a state having multiple sub-states (chase, retreat, heal etc.) through which the enemies would execute a specific action.

Table 1. Encapsulation of the variation in the physical attributes of the simulation with respect to complexities

Attribute	Novice simulation	Professional simulation
Ammunition available	1000	500
Delay between successive shots by the enemy	30 frames	15 frames
Rate of health decrease (enemy) per shot	10	8
Rate of health decrease (player) per shot	1	2

As shown in Table 2, the transition of these states was conditional/probabilistic and depended upon the complexity of the simulation.

Table 2. Encapsulation of the variation in the probabilities of movement with respect to complexities

Movement	Novice simulation	Professional simulation
Towards player	25	75
Towards headquarters	25	75
Random	75	25
Static	75	25
Towards headquarters (Stealth)	30	70
Random	70	30

As shown in Fig. 2, the whole map was divided into sections (layers of the navigation mesh) namely, covered areas, partially open areas and open areas. The cost of moving in these areas was different and the movement was controlled by the navigation mesh. The enemy faction was divided into three groups, with each group consisting of four enemies. One of the enemies was randomly allocated as the leader of the group in every simulation. The other members of the group were programmed to follow the instructions of their leader continually. Three groups were centered at three different headquarters respectively, and one group spawned at a random location in the terrain. These three factions formed the 'assault' group. The assault group did not care about the cost and they moved in all the areas. They had 4 possible destinations: Random, towards player, Towards HQ or static (stay at the same place). The remaining four enemies formed a group called 'stealth'. The stealth enemies tended to walk on layers of the physical terrain that have minimum cost – covered areas. They avoided confrontation with the player and moving out in open areas. They were programmed in such a way that they gave more precedence to destroying the headquarters than attacking the player. Figure 3 shows the state machines for the leader and the other members.

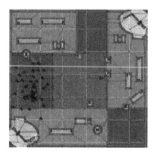

Fig. 2. Picture depicting the different areas in the navigation mesh. Different colors represent different layers. Light blue represents the covered area, with lowest cost. Purple represents partially open areas, with medium cost. Green represents open area, with highest cost. (Color figure online)

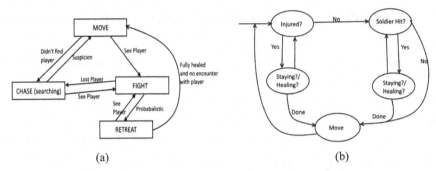

Fig. 3. State machines implemented for the enemies (a) general state of the enemy (b) leader's state

4   Results

We carried out one-way ANOVAs to compare the effect of task difficulty on different the cognitive and behavioral descriptors mentioned above.

4.1   Effects of Professional Training

As shown in Fig. 4(a), the percentage number of enemies killed was significantly higher in the when novice training was given second compared to when it was given first (Novice second: 67.5% > Novice first: 45%; F (1, 28) = 9.62, $p < 0.05$, $\eta2 = 0.27$). The accuracy index was significantly higher when novice training was given second compared to when it was given first (Novice second: 71.5% > Novice first: 53.4%; F (1, 28) = 11.47, $p < 0.01$, $\eta2 = 0.1$) (see Fig. 4(b)). As shown in Fig. 4(c), the self-reported mental demand requirements were significantly higher when novice training was given first compared to when it was given second (Novice first: 6.5 > Novice second: 4.4; F (1, 28) = 18.61, $p < 0.01$, $\eta2 = 0.08$). In addition, as shown in Fig. 4(d), the self-reported effort requirements were significantly higher when novice training was given first compared to when it was given second (Novice first: 5.8 > Novice second: 4.5; F (1, 28) = 6.73, $p < 0.05$, $\eta2 = 0.18$).

4.2   Effects of Novice Training

As shown in Fig. 5(a), the percentage number of enemies killed was significantly higher when professional training was given second compared to when it was given first (Professional second: 56.2% > Professional first: 34%; F (1, 28) = 15.43, $p < 0.01$, $\eta2 = 0.04$). The accuracy index was significantly higher when professional training was given second compared to when it was given first (Professional second: 61.5% > Professional first: 45.5%; F (1, 28) = 10.37, $p < 0.01$, $\eta2 = 0.03$) (see Fig. 5(b)). As shown in Fig. 5(c), the self-reported mental demand requirements were significantly higher when professional training was given first compared to when it was given second (Professional first: 7.9 > Professional second: 6.6; F (1, 28) = 6.89, $p < 0.05$, $\eta2 = 0.1$). In addition, as shown in Fig. 5(d), the self-reported effort requirements were significantly higher when

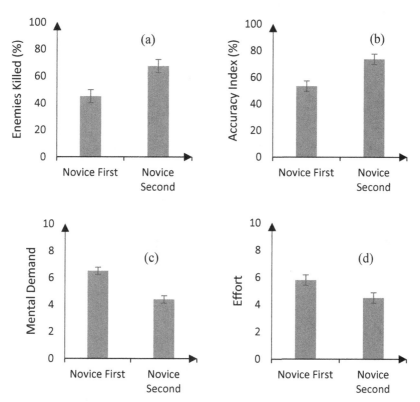

Fig. 4. The averaged performance and cognitive variables across different novice conditions (Novice first and second) (a) Enemies Killed (in %) (b) Accuracy Index (in %) (c) NASA-TLX's mental demand (d) NASA-TLX's self-reported effort requirement

professional training was given first compared to when it was given second (Professional first: 8.5 > Professional second: 7.1; F (1, 28) = 8.75, $p < 0.05$, $\eta2 = 0.36$).

4.3 Inter-training Comparisons

The percentage number of enemies killed was significantly higher when novice training was given second compared to when professional training was given second (Novice second: 67.5% > Professional second: 56.2%; F (1, 28) = 9.85, $p < 0.05$, $\eta2 = 0.22$). The accuracy index was significantly higher when novice training was given second compared to when professional training was given second (Novice second: 71.5% > Professional second: 61.5%; F (1, 28) = 13.82, $p < 0.05$, $\eta2 = 0.07$). This result could simply be because novice training was simpler compared to the professional training. The self-reported mental demand requirements were significantly higher when professional training was given second compared to when novice training was given second (Professional second: 6.6 > Novice second: 4.4; F (1, 28) = 17.44, $p < 0.01$, $\eta2 = 0.01$). In addition, the self-reported effort requirements were significantly higher when professional training was given second compared to when novice training was given

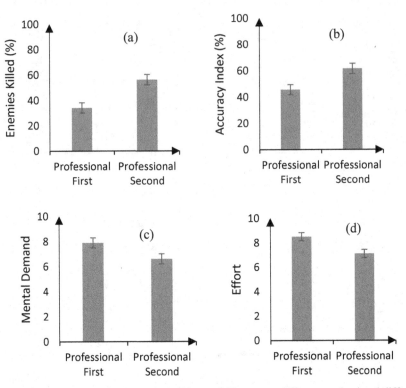

Fig. 5. The averaged performance and cognitive variables across different professional difficulty conditions (Professional first and second) (a) Enemies Killed (in %) (b) Accuracy Index (in %) (c) NASA-TLX's mental demand (d) NASA-TLX's self-reported effort requirement

second (Professional second: 7.1 > Novice second: 4.5; $F(1, 28) = 19.56, p < 0.01, \eta 2 = 0.03$).

5   Discussion

In this experiment, we evaluated the cognitive and behavioral implications of Virtual Reality training under varying degrees of task difficulty in a complex search and shoot scenario. Results suggested that the participants fared better in the Novice/ Professional second design compared to the Novice/Professional first design. These results seemed to comply with [9, 11], who had argued that difficult objectives/goals during training can serve as a stimulus for strategy development. IBLT [14] also provides an explanation for our results, where the professional training provided more information first-hand which also led to the formation of more instances in the participant's memory. The formation of these instances conceivably led to better performance in the transfer simulation. As shown in Figs. 4(c), (d), (c), and (d), the self-reported NASA-TLX measures also reported higher requirements of mental demand and effort after novice training compared to after professional training. These results also find its basis on IBLT [14], where more information availability and compartmentalization of cognitive resources in professional training compared to novice training led to an increase the information-processing requirement for storing these instances created in memory. The main implications of our results suggest that in virtual reality simulations in professional (difficult) scenarios/conditions provide better transfer compared to novice (easy) scenarios/conditions. Thus, training soldiers involving professional scenarios is likely to yield better performance in the real world. As part of future research, we plan to extend our current investigations to design a human performance modeling framework to achieve optimal transfer of cognitive skills in these interfaces. Also, we plan to compare human performance due to different training interventions in 2D indirect visual displays.

6 Conclusion

This experiment proved that training the individual in a difficult condition leads to efficient performance even in a dynamic, multi-dimensional and complex task. We expect to use the conclusions from this experiment as a means of creating effective training environments for military personnel using virtual-reality interfaces.

Acknowledgments. This research was supported by a grant from Defence Research and Development Organization (DRDO) titled "Development of a human performance modeling framework for visual cognitive enhancement in IVD, VR and AR paradigms" (IITM/DRDO-CARS/VD/110) to Varun Dutt.

References

1. Ministry of Defence: National Security Through Technology: Technology, Equipment, and Support for UK Defence and Security, February 2012. https://www.gov.uk/government/uploads/system/uploads/attachment_data/file/27390/cm8278.pdf
2. ter Haar, R.: Virtual reality in the military: present and future. In: 3rd Twente Student Conference on IT (2005)
3. McCollum, C., Barba, C., Santarelli, T., Deaton, J.: Applying a cognitive architecture to control of virtual non-player characters. In: Proceedings of the 2004 Winter Simulation Conference, vol. 1. IEEE, December 2004
4. Gilbert, S., Boonsuk, W., Kelly, J.W.: Virtual displays for 360-degree video. In: Human Vision and Electronic Imaging XVII, vol. 8291, p. 82911L. International Society for Optics and Photonics, February 2012
5. Cosenzo, K., et al.: Impact of 360 Sensor Information on Vehicle Commander Performance. Army Research Laboratory, Aberdeen Proving Ground, MD ARL-TR-5471 (2011)
6. Freina, L., Canessa, A.: Immersive vs desktop virtual reality in game based learning. In: European Conference on Games Based Learning, p. 195. Academic Conferences International Limited, October 2015
7. McCarley, J.S., Wickens, C.D.: Human factors implications of UAVs in the National Airspace (2005)
8. Rao, A.K., Pramod, B.S., Chandra, S., Dutt, V.: Influence of indirect vision and virtual reality training under varying manned/unmanned interfaces in a complex search-and-shoot simulation. In: Cassenti, D.N. (ed.) AHFE 2018. AISC, vol. 780, pp. 225–235. Springer, Cham (2019). https://doi.org/10.1007/978-3-319-94223-0_21

9. Pomplun, M., Garaas, T.W., Carrasco, M.: The effects of task difficulty on visual search strategy in virtual 3D displays. J. Vis. **13**(3), 24 (2013)

10. Williams, D.E., Reingold, E.M., Moscovitch, M., Behrmann, M.: Patterns of eye movements during parallel and serial visual search tasks. Can. J. Exp. Psychol./Revue canadienne de psychologie expérimentale **51**(2), 151 (1997)

11. Madhavan, P., Gonzalez, C., Lacson, F.C.: Differential base rate training influences detection of novel targets in a complex visual inspection task. In: Proceedings of the Human Factors and Ergonomics Society Annual Meeting, vol. 51, no. 4, pp. 392–396. SAGE Publications, Los Angeles, October 2007

12. Mane, A., Wickens, C.D.: The effects of task difficulty and workload on training. In: Proceedings of the Human Factors Society Annual Meeting, vol. 30, no. 11, pp. 1124–1127. SAGE Publications, Los Angeles, September 1986

13. Gopher, D., North, R.A.: Manipulating the conditions of training in time-sharing performance. Hum. Factors **19**(6), 583–593 (1977)

14. Gonzalez, C., Dutt, V.: Instance-based learning: integrating sampling and repeated decisions from experience. Psychol. Rev. **118**(4), 523 (2011)

15. Srivastava, P., Rimzhim, A., Vijay, P., Singh, S., Chandra, S.: Desktop VR is better than non-ambulatory HMD VR for spatial learning. Front. Robot. AI **6**, 50 (2019)

16. Murcia-López, M., Steed, A.: The effect of environmental features, self-avatar, and immersion on object location memory in virtual environments. Front. ICT **3**, 24 (2016)

17. Pyc, M.A., Rawson, K.A.: Testing the retrieval effort hypothesis: does greater difficulty correctly recalling information lead to higher levels of memory? J. Mem. Lang. **60**(4), 437–447 (2009)

18. Creighton, R.H.: Unity 3D Game Development by Example: A Seat-of-Your-Pants Manual for Building Fun, Groovy Little Games Quickly. Packt Publishing Ltd., Birmingham (2010)

19. Roosendaal, T., Selleri, S.: The Official Blender 2.3 Guide: Free 3D Creation Suite for Modeling, Animation, and Rendering. No Starch Press, San Francisco (2004)

20. Hart, S.G., Staveland, L.E.: Development of NASA-TLX (task load index): results of empirical and theoretical research. In: Hancock, P.A., Meshkati, N. (eds.) Advances in Psychology, vol. 52, pp. 139–183. North-Holland, Amsterdam (1988)

Glyph Reader App: Multisensory Stimulation Through ICT to Intervene Literacy Disorders in the Classroom

Paola Ariza-Colpas[1]([✉]), Alexandra Leon-Jacobus[2], Sandra De-la-Hoz[1],
Marlon Piñeres-Melo[3], Hilda Guerrero-Cuentas[1], Mercedes Consuegra-Bernal[4],
Jorge Díaz-Martinez[1], Roberto Cesar Morales-Ortega[1],
and Carlos Andrés Collazos Morales[5]

[1] Universidad de la Costa, CUC, Barranquilla, Colombia
{pariza1,sdelahoz2,hguerrero,jdiaz5,rmorales1}@cuc.edu.co
[2] Universidad Metropolitana de Barranquilla, Barranquilla, Colombia
aleonj@unimetro.edu.co
[3] Universidad del Norte, Barranquilla, Colombia
pineresm@uninorte.edu.co
[4] Universidad Simon Bolivar, Barranquilla, Colombia
mconsuegra@unisimon.edu.co
[5] Universidad Manuela Beltran, Bogotá, Colombia
carlos.collazos@docentes.umb.edu.co

Abstract. This article shows the experience in the implementation of a tool called Glyph Reader, which is an application that has two interfaces, Web and Mobile and that responds to the need for an educational and interactive resource whose main objective is the Multisensory stimulation for literacy training in a population with cognitive disabilities and/specific learning disorder. The design of the activities that this application has is based on the theoretical model of multisensory stimulation Orton Gillingham, which seeks the development of basic skills for decoding isolated words based on a phonetic - graphic analysis of them. The techniques within this model use the basic concepts of intersensory integration of simultaneous visual-auditory-kinesthetic- tactile differentiation (VAKT), to which the Glyph Reader application takes full advantage, by including graphic phonetic recognition and training activities of syllables/words (exercises with symphons and exercises with combinations of consonants or working syllables), which pass from basic levels to complex levels of decoding, necessary for the development of literacy skills. The study sample for software validation is 250 students from the Eustorgio Salgar educational institution, in the municipality of Puerto Colombia, in the department of Atlántico - Colombia

Keywords: Multisensory stimulation · Phonetic recognition · Technological platform

1 Introduction

Taking into account the needs that exist from the education sector to guarantee the entry to quality education as a fundamental right of persons with disabilities; and with it, the

U. S. Tiwary and S. Chaudhury (Eds.): IHCI 2019, LNCS 11886, pp. 259–269, 2020.
https://doi.org/10.1007/978-3-030-44689-5_23

deployment for the creation and offer of pedagogical support services for the inclusion of this population, the present project is formulated and implemented which results in the design of the Glyph Reader tool, which seeks to enhance and/or develop literacy competence under a multisensory stimulation approach in population with intellectual disability and/or specific learning problems.

According to the WHO and the World Bank (2011) [1], it is estimated that more than one billion people live with disabilities (around 15% of the world population), and in the specific case of the child population (0–14 years), Statistics reveal that an average of 95 million children (5.1%) present some type of disability, being vulnerable populations, those with low economic income or belonging to ethnic minority groups, who present a significantly higher risk. In the child population, it has been identified that one of the disabilities that have the greatest impact on functionality and school adaptation is Cognitive or Intellectual Disability, with a prevalence of 1% according to the DSM-V Diagnostic and Statistical Manual of Mental Disorders [2], percentage that varies according to age, as well as the level of commitment (intellectual disability of severe functional commitment occurs in 6 of every 1000 people).

Intellectual Disability, is defined by the American Psychiatric Association [2] as the deficit in intellectual functioning (reasoning, problem solving, planning, thinking, academic learning, among others) confirmed through standardized clinical evaluations or intelligence tests; also involving deficits in adaptive functioning that result in the non-achievement of social and cultural standards for personal independence and social responsibility. According to Márquez [3], social and educational intervention implies recognizing that intellectual disability impacts on family and community adaptation in which the child develops, which is why timely school insertion must be taken as a priority goal associated with the improvement in the quality of life, since it tends to reduce barriers to learning, participation, and promote social inclusion.

Along with intellectual disability, other difficulties present in early childhood are related to learning problems in reading and writing. Hallahan, Kauffman and Lloyd [4], point out that the reading disorder (dyslexia) is frequently associated with the disorder of calculus and written expression, being relatively rare to find any of these disorders in the absence of it (the reading one). Reading difficulties are very important to evaluate, since the disparity related to the development of these competences leads to significant implications in the success of other academic areas [22, 23]. They also highlight the fact that skills in this area are considered the most valuable for culture. That is why, in the elementary grades, much of the curriculum focuses on the acquisition of reading ability and in the secondary grades it is the main vehicle for the presentation of information in the contents of different subjects.

In a population without some type of disability, McLoughlin and Lewis [5], point out that, by the third grade, students with reading problems begin to fall behind their classmates in other subjects. Thus, according to Stanovich [6], one of the consequences of this is that these subjects have less access than their peers to sources of information; In this way, while children who have already appropriated reading obtain new information through the use of their ability, those with difficulties continue to fall behind in other areas, differentiating themselves from their peers in the degree of knowledge and information they provide. the work material for the development of your intelligence [24, 25].

One of the strategies to facilitate the social inclusion of people with disabilities and reduce the inequality gap is through technology. According to Aleída Fernández and Yinzú Nairouz "ICT can provide socially rich environments so that people with disabilities can make their disadvantage visible through dialogue and participation in different contexts on and off the Internet." In view of this, the need for actions that seek to know the current situation regarding the availability of enriched projects, tools and inclusive spaces through the use of IT becomes visible [28, 29].

2   Brief Review of Literature

Doval [7], performs algorithm analysis in the English language. In this work we experiment with a wide range of phonetic algorithms for the English language. The objective of this experiment is to identify different phonetic variables within the framework of the analyzed texts that serve as support for the automatic analysis. Ram [8], analyse the difficulty of detecting voice expressions, based on large voice files. The author performs an exploration based on deep neural networks to be able to analyze the different phonetic spaces and subspaces for the extraction of patterns in the different queries that are required to be performed on the audio that is analyzed.

Kos [9], developed a support platform for multilingual analysis for the detection of different phonetic patterns in speech action. The main objective of this work is to achieve a rapid adaptation of the Slovenian base language to different languages through voice processing. Ghosh [10], focuses its work on the interpretation of the script Devanagari, which is considered as one of the most popular scripts in all of India, for the analysis of the respective patterns and symbols in the application Hidden Markov Models (HMM) has been used for detection Automatic of the different symbols. Cooper [11], studies the formal analysis of the analysis of Dutch English audios, to determine the number of coincidences in the language using neural networks and computational intelligence.

At present, many focused applications have been developed whose purpose is to take advantage of the use of ICT, as support for the improvement of intellectual disability, which are related below:

- Read it easy [12]: This application was developed by the Spanish Confederation of people with intellectual or developing disabilities FEAPS, which aims to support the learning process Reading, aimed primarily at people with intellectual or developmental disabilities or those who have difficulty reading. This project has allowed the adaptation of literary works in an easier to understand format for adaptation to reading and understanding. In the same way this application also has application in the learning of the Spanish language for immigrants to this country.
- Special Words [13]: This application allows you to support children with special educational characteristics in the recognition and learning of words. This process is achieved by matching each of the words with their drawings and sounds. This application includes the use of 96 base words, which can be customized, in the same way children can organize these words and also add others that are related to your daily life as household items, family members, friends among others. The application manages difficulty levels to be able to evaluate the level of learning of users.

– Special Numbers [14]: The main objective of this application is and develop in young people and children with learning difficulties, learning numbers, working mainly in the areas of: Numerical comparison, Ordering and counting. For the development of this application, there was the participation of parents, children, teachers and psychopedagogues, taking into account the way in which children of different ages learn acquires mathematical skills and abilities.

– Series 1 [15]: Focused on children from 3 and a half years, the Series 1 application allows the ordering of objects taking into account their size, color, shape, among other features. This apps are
It is made up of a set of several boards in which children manage to visualize different objects that they must order in accordance with the request made by the software.

– KIMI [16]: This application was developed by the University of Deusto and the Down Syndrome Foundation of the Basque Country, designed to promote healthy living habits in children and adolescents with intellectual and intellectual disabilities. The application achieves this goal through its central character Kiwi, an alien who arrives on earth and needs to learn the customs of the planet, through interactive and intuitive approaches.

– Drawing [17]: This application was developed by Global Labs, whose objective is the stimulation of space construction in children. Through colorful and entertaining scenarios, children must locate objects that disappear from the stage. The child makes the selection and change of position of the objects from the initial position to the final position, when he manages to position the objects correctly he advances in the application levels.

– Grupolandia [18]: This application was developed by ASDRA and InfinixSoft, which makes use of elements such as: school supplies, fruits, toys to stimulate the classification of different collections of objects.

– Tuli Emotions [19]: This application was developed by IBM Argentina, as a result of the On Demand Community research program of Corporate Citizenship. The end user was ASDRA (Down Syndrome Association of the Argentine Republic), which takes a set of situations from everyday life to stimulate the emotional area of children. In this application children select from a set of faces, which represents the emotion that is described in the scenario shown, when he manages to locate the emotion with the real stage he receives applause and chooses to continue with the next scenario. When the face does not match the selected scenario, it returns to the wrong starting position.

3   Methodology

For the development of the components of the Glyph Reader application, taking into account the levels of scope, it was necessary to know the characteristics at the level of learning and acquisition of the literacy competence, of those who will constitute the sample of users of the initial version of the tool [12, 13]. The importance of having a description of the levels of acquisition of literacy in the population to be involved, as well as coming into contact with teachers and parents or guardians to publicize.

This type of iterative development allowed the feedback of the research and development process in its early stages, so that from the beginning data and information can be obtained that enable improvements to the tool designed and guarantee the success of the final product.

The characterization of the population is carried out in the facilities of the San Juan Bosco Educational Institution, which is of an official nature and is located in the municipality of Sabanagrande - Atlántico. The Institution was chosen as the basis for the development of the project, in the initial phase of characterization, since it has an included population that presents cognitive disability, in addition, it reports high rates of low academic performance and also presents a history of low performance on the Knowledge tests [14, 15]. At present, the Institution has an approximate population of 1500 students, which are distributed in the levels of Preschool, Basic Primary, Basic Secondary and Academic Average.

3.1 Population and Sampling

The population under study are students between the ages of 7 and 14, who have difficulties in learning to read or write or who have a diagnosis of cognitive impairment of the level of mild commitment.

Likewise, the choice of the sample is given by expert criteria, of a non-probabilistic type, because it is made up of elements of a population that are chosen at random [16, 17]. Therefore, the following phases were developed, within the framework of the methodology:

1. Awareness raising of the project to the community of Managers and Teachers.
2. Information Request for students who meet the inclusion characteristics.
3. Meeting with parents and students who will be part of the study.
4. Informed consent signatures.
5. Request for information on personal data and academic performance.
6. Diligence of the characterization instrument by the teacher community (group director).
7. Tabulation of data and interpretation of results.
8. Results and preliminary report.

3.2 Instruments

To obtain general information of the level in which the students are, in addition to consulting notes of the first period for the subjects of mathematics and Spanish, the Questionnaire for the Assessment of Learning Problems (CEPA) was applied, which the director of grade for each student reported [26, 27]. CEPA is "an instrument designed to be used by the teacher himself in the classroom. Its objective is to provide an instrument for the evaluation of the learning difficulties that students have in different areas, within the context of their own course. This facilitates the detection of children who are more likely to have learning difficulties and their timely referral for a diagnosis" [18, 19].

CEPA comprises 39 items that are grouped into 8 frequent areas of learning difficulties, is designed to evaluate processes Cognitive and language. The evaluation areas are as follows [20, 21]:

- Receipt of information: Includes 4 items and has as purpose to evaluate the child's ability to understand and retain oral information.
- Oral language expression: Evaluates the use of expressive language.
- Attention concentration and memory: Estimate the capacity of the child to attend in class the concepts given by the teacher. Its concentration and working memory.
- Reading: Evaluate confusion when reading letters and words, their type of reading and comprehension.
- Writing: The achievements in copying, dictation, calligraphy and spelling are evaluated.
- Mathematics: Measures knowledge of ordinal numbers, cardinal, the ability to perform arithmetic operations and the understanding its meaning.
- Global Evaluation: It is the appreciation of the intellectual capacity of the child, according to the teacher's criteria.

Based on each component, the teacher must give a score for each statement, which is expressed as follows:

0. Strongly Disagree
1. Partially Disagree
2. Partially agree
3. Totally agree

4   Applications Features

Glyph Reader is an application based on multisensory analysis that allows the strengthening of reading and writing skills in children between the ages of 7 and 13, through a guided interface of three types of activities: Drawing, Writing and Choosing (Figs. 1 and 2).

Fig. 1. Principal screen of Glyph Reader   **Fig. 2.** Selection roles in Glyph Reader

When the user selects the draw option, it allows him to make strokes freely so that he can draw on the screen in accordance with the phoneme he has heard. By means of the write option, the syllable that you wish to reinforce can be redrawn based on the audio that is supplied by the software (Figs. 3 and 4).

Fig. 3. Writting in Glyph Reader   **Fig. 4.** Chossing in Glyph Reader

In the option you choose, given an audio supplied by the system the user must select the correct writing option. All these types of exercises strengthen the listening and writing skills of children with cognitive and intellectual disabilities to improve reading-writing skills.

5   Results

Once the students who meet the inclusion criteria (age between 7 and 14 years old, poor performance in literacy tasks, identify a diagnosis of mild cognitive impairment or difficulty in reading or writing learning) have been identified and whose Parents or guardians sign the informed consent, an initial group of 59 participants is obtained, from which the following information is obtained (Table 1):

Table 1. Age table

Age	Frequency	Percentage
7	1	1,7%
8	13	22,0%
9	14	23,7%
10	15	25,4%
11	7	11,9%
12	7	11,9%
13	2	3,4%
Grand total	59	100%

Of the children who will be part of the research, it is found that the highest percentage is in ages ranging from 8 to 12 years, children who are behind in relation to their peers and who, according to the assessment of their teachers, present problems in the correct writing of words and syllables, incurring errors that are not admissible for their age and level of education. A homogeneous distribution of boys and girls is determined (Tables 2 and 3).

Table 2. Sex table

Sex	Frequency	Percentage
Male	29	49,2%
Female	30	50,8%
Grand total	59	100%

As shown in the Table 4, there is a greater participation of third grade students. This occurs, because in those grades the SABER tests are presented, since there is special interest on the part of the directive and teaching community to know what happens to those children who are behind in their reading-writing process despite their age and level. Likewise, it is desired to see the effectiveness of the tool in third grade children, since short-term strategies can be implemented to enhance their reading skills. On the other hand, there are children from the Acceleration and Compass programs, which include students with learning difficulties due to external variables (lack of stimulation or support at home, etc.) and/or diagnoses of cognitive disability or problems of learning.

As can be seen in the Table 4, of the 59 students belonging to the study sample, the majority (59%), that is; 35 students present a basic academic performance and 23 students (39%) a low performance. Only 1 student shows high performance according to their grades.

Table 3. Grade table

Age	Frequency	Percentage
1° B	5	8,5%
3° A	10	16,9%
3° B	13	22,0%
3° C	13	22,0%
3° D	4	6,8%
4° A	3	5,1%
5° A	2	3,4%
5° B	4	6,8%
Acceleration	2	3,4%
Compass	3	5,1%
Grand total	59	100%

Table 4. Math performance table

Math performance	Frequency	Percentage
High performance	1	1,7%
Low performance	23	39,0%
Basic performance	35	59,3%
Grand total	59	100%

Table 5. Spanish performance table

Spanish performance	Frequency	Percentage
Low performance	17	28.8%
Basic performance	42	71.2%
Grand total	59	100%

With regard to the basic Spanish subject, the students who are part of the study have a basic or low performance. As teachers were requested and as expected, students who will present problems such as confusing letters, literacy problems, difficulties in separating words, among others; They are students who have poor performance in basic subjects such as mathematics and Spanish. After the use of the software, the following improvements could be identified in the students (Tables 5, 6 and 7).

Table 7. Spanish performance table after software implemented

Spanish performance	Frequency	Percentage
Low performance	27	45,7%
Basic performance	32	54,3%
Grand total	59	100%

Table 6. Math performance Table after software implemented

Math performance	Frequency	Percentage
High performance	24	40,67%
Basic performance	35	59,3%
Grand total	59	100%

6 Conclusions

In the development of the characterization process significant advances are achieved, among which the following stand out:

1. It is possible to establish commitments on the part of the educational Institution in development of the project, as well as of the educational community and parents.
2. It is possible to present the demo version of the application to teachers, parents and students, so that they give their feedback on the tool, so that with their contributions they can go to work and improving the design and development of it.
3. It is observed in the responses to the CEPA questionnaire that teachers report difficulties in reading in children selected at the level of syllable reading, word reading and reading sentences and also consider that children have a low level of development of their processes of learning, regarding what is expected for their age and level of education.

For the purpose of giving greater impact to the final investigation, a second part of the characterization will continue, including the application of an intelligence scale (Weschler Intelligence Scale - WISC-IV) and the ENI- reading and writing subtests. 2 (Child Neuropsychological Evaluation 2), instruments that will be provided by the Psychobiology Laboratory of the Faculty of Psychology of the University of Costa CUC.

Acknowledgment. To the company BlazingSoft for the development of the software GLYPH READER.

References

1. Dyachenko, Organización Mundial de la Salud. Informe Mundial Sobre la Discapacidad. https://www.who.int/disabilities/world_report/2011/summary_es.pdf
2. American Psychiatric Association: Guía de consulta de los Criterios Diagnósticos del DSM-5™. http://www.eafit.edu.co/ninos/reddelaspreguntas/Documents/dsm-v-guia-consulta-manual-diagnostico-estadistico-trastornos-mentales.pdf
3. Marquez, J.A., Ramírez García, J.I.: Family caregivers' narratives of mental health treatment usage processes by their Latino adult relatives with serious and persistent mental illness. J. Fam. Psychol. **27**(3), 398–408 (2013). https://doi.org/10.1037/a0032868
4. Kauffman, J.M.: Research to practice issues. Behav. Disord. **22**(1), 55–60 (1996)
5. McLoughlin, J.A., Lewis, R.B.: Assessing Special Students. Macmillan College, New York (1994)
6. Stanovich, K.E.: How to Think Straight About Psychology (1986)
7. Doval, Y., Vilares, M., Vilares, J.: On the performance of phonetic algorithms in microtext normalization. Expert Syst. Appl. **113**, 213–222 (2018)
8. Ram, D., Asaei, A., Bourlard, H.: Phonetic subspace features for improved query by example spoken term detection. Speech Commun. **103**, 27–36 (2018)
9. Kos, M., Rojc, M., Žgank, A., Kačič, Z., Vlaj, D.: A speech-based distributed architecture platform for an intelligent ambience. Comput. Electr. Eng. **71**, 818–832 (2018)

10. Ghosh, R., Vamshi, C., Kumar, P.: RNN based online handwritten word recognition in Devanagari and Bengali scripts using horizontal zoning. Pattern Recogn. **92**, 203–218 (2019)

11. Cooper, A., Bradlow, A.: Training-induced pattern-specific phonetic adjustments by first and second language listeners. J. Phon. **68**, 32–49 (2018)

12. Read it easy. http://www.feaps.org/

13. Special Words. https://itunes.apple.com/co/app/palabras-especiales/id451723454?mt=8

14. Special Numbers. https://itunes.apple.com/es/app/numeros-especiales/id531381202?mt=8

15. Series 1. https://itunes.apple.com/es/app/series-1/id501937796?mt=8

16. Kimi. https://itunes.apple.com/es/app/Kimi/id501937796?mt=8

17. Drawing.   https://play.google.com/store/apps/details?id=com.globant.labs.dibugrama.android

18. GrupoLandia. https://play.google.com/store/apps/details?id=com.infinixsoft.asdragrupos

19. Tuli Emotions. https://play.google.com/store/apps/details?id=com.tuli2.main

20. Briones, C.: "Con la tradición de todas las generaciones pasadas gravitando sobre lamente de los vivos": usos del pasado e invención de la tradición (2003)

21. Bravo-Villasante, C., Muñoz, S.: Antología de la literatura infantil española (1979)

22. Arango, L.M., et al.: Educación física y desarrollo humano integral (2002)

23. De-La-Hoz-Franco, E., Ariza-Colpas, P., Quero, J.M., Espinilla, M.: Sensor-based datasets for human activity recognition–a systematic review of literature. IEEE Access **6**, 59192–59210 (2018)

24. Ariza, P., Pineres, M., Santiago, L., Mercado, N., De la Hoz, A.: Implementation of MOPROSOFT level I and II in software development companies in the Colombian Caribbean, a commitment to the software product quality region. In: 2014 IEEE Central America and Panama Convention (CONCAPAN XXXIV), pp. 1–5. IEEE, November 2014

25. Ariza-Colpas, P., Oviedo-Carrascal, A.I., De-la-hoz-Franco, E.: Using K-means algorithm for description analysis of text in RSS news format. In: Tan, Y., Shi, Y. (eds.) DMBD 2019. CCIS, vol. 1071, pp. 162–169. Springer, Singapore (2019). https://doi.org/10.1007/978-981-32-9563-6_17

26. Piñeres-Melo, M.A., Ariza-Colpas, P.P., Nieto-Bernal, W., Morales-Ortega, R.: SSwWS: structural model of information architecture. In: Tan, Y., Shi, Y., Niu, B. (eds.) ICSI 2019. LNCS, vol. 11656, pp. 400–410. Springer, Cham (2019). https://doi.org/10.1007/978-3-030-26354-6_40

27. Ariza-Colpas, P.P., Piñeres-Melo, M.A., Nieto-Bernal, W., Morales-Ortega, R.: WSIA: web ontological search engine based on smart agents applied to scientific articles. In: Tan, Y., Shi, Y., Niu, B. (eds.) ICSI 2019. LNCS, vol. 11656, pp. 338–347. Springer, Cham (2019). https://doi.org/10.1007/978-3-030-26354-6_34

28. Ariza-Colpas, P., et al.: Enkephalon - technological platform to support the diagnosis of alzheimer's disease through the analysis of resonance images using data mining techniques. In: Tan, Y., Shi, Y., Niu, B. (eds.) ICSI 2019. LNCS, vol. 11656, pp. 211–220. Springer, Cham (2019). https://doi.org/10.1007/978-3-030-26354-6_21

29. Ariza-Colpas, P., Morales-Ortega, R., Piñeres-Melo, M., De la Hoz-Franco, E., Echeverri-Ocampo, I., Salas-Navarro, K.: Parkinson disease analysis using supervised and unsupervised techniques. In: Tan, Y., Shi, Y., Niu, B. (eds.) ICSI 2019. LNCS, vol. 11656, pp. 191–199. Springer, Cham (2019). https://doi.org/10.1007/978-3-030-26354-6_19

Cyclon Language First Grade App: Technological Platform to Support the Construction of Citizen and Democratic Culture of Science, Technology and Innovation in Children and Youth Groups

Paola Ariza-Colpas[1]([✉]), Belina Herrera-Tapias[1], Marlon Piñeres-Melo[2], Hilda Guerrero-Cuentas[1], Mercedes Consuegra-Bernal[3], Ethel De-la-Hoz Valdiris[1], Carlos Andrés Collazos Morales[4], and Roberto Cesar Morales-Ortega[1]

[1] Universidad de la Costa, CUC, Barranquilla, Colombia
{pariza1,bherrera3,hguerrero,edelahoz3,rmorales1}@cuc.edu.co
[2] Universidad del Norte, Barranquilla, Colombia
pineresm@uninorte.edu.co
[3] Universidad Simon Bolivar, Barranquilla, Colombia
mconsuegra@unisimon.edu.co
[4] Universidad Manuela Beltran, Bogotá, Colombia
carlos.collazos@docentes.umb.edu.co

Abstract. This article shows the construction of software applications Cyclon Language First Grade App, like a strategy in which communities of practice, learning, knowledge, innovation and transformation are generated, understood as a transversal process, where collaborative, problematizing learning is encouraged, by critical inquiry, permanent interaction, cultural negotiations and the dialogue of knowledge, typical of the pedagogical proposal of the Ondas program. It is summarized in the following aspects: "Building an identity that incorporates the recognition of science and technology as a constituent element of everyday culture both in individuals and in the communities and institutions of which they are part, involving various sectors of society: productive, social, political, state and in the various territorial areas: local, departmental and national. Development of forms of organization oriented to the appropriation of values that recognize a cultural identity around science and technology in the aspects mentioned in the previous point. This implies models of participation, social mobilization and public recognition of scientific and technological activity. On the other hand, the incorporation of the research activity in the elementary and middle school involves the development of national, departmental and loçal financing mechanisms; in such a way that children and young people can develop their abilities and talents in a favorable environment of both social recognition and economic conditions. Development of a methodological strategy supported by ICT that helps the Colombian population to recognize and apply both individually and collectively, science and technology through research activities designed according to the characteristics of the scientific method. "The appropriation of ICTs as a constitutive part of the citizen and democratic culture of the CT + I and the construction of virtual reality as central to the process of knowledge democratization.

U. S. Tiwary and S. Chaudhury (Eds.): HCI 2019, LNCS 11886, pp. 270–280, 2020.
https://doi.org/10.1007/978-3-030-44689-5_24

Keywords: Citizen and democratic culture · Technology and innovation ·
Technological platform · Spanish language learning

1 Introduction

The appropriation of this culture implies the development of scientific, technological, innovation, social, cognitive and communicative abilities, skills and competences and the capacities to inquire and observe, which are consolidated as children, girls and young people receive adequate guidance in approaching their problems, through daily work in the different spaces of socialization; In this sense, educational institutions play the main role: preparing their teachers in methodologies that favor such appropriation, and seeking alliances with academic and non-academic entities that conduct research [1].

This project was a special interest in the construction of a citizen culture in science, technology and innovation, promoting in the students and teachers of the Department the formation of communities and the realization of research that seeks to solve the problems of their environment and build capacities. To move in a world that makes its reorganization from the new processes of knowledge and knowledge, founded on the CT + I. The constituent elements of this culture are [2, 3]:

- Demystification of science, its activities and products to be used in everyday life and in solving problems.
- Democratization of knowledge and knowledge guaranteeing its appropriation, production, use, conversion, storage and transfer systems in all sectors of society.
- The capacity of judgment and criticism about its logic, its uses and consequences.
- The skills, abilities and competences derived from these new realities (technological, scientific, cognitive, social, valued, communicative, propositive and innovation).
- The skills, abilities and knowledge for research.
- Collaborative learning and the ability to relate to organizational systems in communities of knowledge and knowledge, networks and lines of research.
- The incorporation in the pedagogical and investigative processes of information and communication technologies.
- The ability to ask, raise problems and give them creative solutions through the development of inquiry processes.
- The development of creativity through actions that lead to innovations. The ability to change in the midst of change.

2 Pedagogies Focused on Research

A methodological line that tries to give way to the contemporary debate about science and its impact on education, uses research to boost school processes, and generate methodological alternatives to build a school close to the configuration of a scientific spirit. In this line, research is understood as the basic tool of knowledge production, and it is the support to introduce children and young people in the path of critical thinking, which facilitates the learning that corresponds to their age group [4].

Various methodological proposals arise, coming from different latitudes. His concern is not only research, but also the act of teaching and learning, and therefore, the required teacher profile and its role in the face of knowledge. Likewise, the required efforts of this professional shape a different institutional framework, consistent with the methodological commitment to the development of school life. Teaching focused on research takes multiple paths and places particular emphasis, not all convergent or complementary.

The pedagogical strategy focused on research is, according to the policy of training human resources and social appropriation of scientific and technological knowledge of Colciencias, the fundamental axis to foster a citizen culture of CT + I in children, girls and boys. young Colombians, because they recognize in them their ability to explore, observe, ask about their environments, their needs and their problems; through project design [5, 6].

Therefore, it is necessary to recognize the role that research activity can play in the modern world, which means a displacement of an activity that was always in the adult world and today clearly appears in that of Colombian children and young people, as a practice transferred to multiple spheres of society, and very particularly, within children's and youth cultures, marking their personal developments, their socialization and their learning [23, 24].

In this sense, today a network is built between research and the world of children and youth that makes it impossible for these practices not to be exogenous to these cultures, but that "it is recognized as an activity of the human being, possible to develop in all areas of knowledge and with very young children. It is also understood as a process of deciphering the human condition from the life experience of this population in school, family and community contexts" [7].

Given that these processes are necessary in the practices, not only academic and scientific, but also of everyday life, it is necessary to "develop a stage of sensitization and induction to infants and young people about the importance of research as the fundamental axis of their training process". In this context, and from the perspective of the Waves program, developing research in school implies [25, 26]:

1. Understand that research questions become permanent and arise from the interest, initiatives and concerns of educational actors.
2. Assume that research must produce various benefits for children and young people: some, in relation to the construction of scientific knowledge and therefore, with the advances of CT + I; others, with the development of research skills and abilities of the subjects.
3. To promote, from a very early age, the cognitive, communicative and social capacities in children, with which they could explore the academic world that is presented to them, towards the search for a meaning for their life.
4. Build meaningful experiences for boys, girls and young people, through pedagogical strategies that link them as central actors in the process.

In this context, the Project, conceives research as a process of deciphering reality based on questions and problems identified by children and young people, is the specific mechanism on which the program methodology is built.

Through it it is feasible to meet expectations of social and personal order for those who undertake this task, because through it it is possible "on the one hand, find real solutions to social problems and on the other [can be assumed by] restless individuals, with initiative as a life perspective. In this sense, it does not only imply the construction of knowledge, but also the transformation of social realities [8, 9].

3   Brief Review of Literature

Performing a review of the literature we can highlight the following works that have been developed in order to enhance brain activities, which are detailed below:

Brain Training (that is, improving, rehabilitating, or simply maintaining cognitive function through deliberation cognitive exercise) is rapidly growing in popularity, but remains highly controversial. Among the biggest problems in current research is the lack of a measure of participants' expectations, which can influence the degree to which they improve with training. This research is based on a questionnaire to measure the perceived effectiveness of brain training software. Initially in this investigation the expectations of the participants were measured at the beginning of the study, and then at the end of the investigation, the effectiveness of the process is measured, whether high or low. Based on the knowledge they have collected from advertising and other real-world sources, people are relatively optimistic about brain training. However, short messages can influence expectations about reported brain training results: reading a brief positive message can increase reported optimism, while reading a short negative message can decrease it. Older adults seem more optimistic about brain formation than young adults, especially when they report being well informed about the formation of the brain and computers. These data indicate that perceptions of brain formation are malleable to at least some extent, and may vary depending on age and other factors. The questionnaire can serve as a simple tool, which is easily incorporated to assess the apparent validity of brain training interventions and create a covariate to account for expectations in statistical analyzes [10].

These types of solutions have also been implemented to other sectors such as medicine, [11] working a memory loss is common in patients with heart failure (HF), but there are few interventions that have been proven to counteract it. The objective of this study was to evaluate the effectiveness of a cognitive training intervention, Brain Fitness, to improve memory, brain-derived neurotropic factor (BDNF) levels, working from memory, processing speed, executive function, activities instrumental of daily life, mobility, depressive symptoms and quality related to the health of life.

Also in depression analysis Cognitive training (CCT) through computerized paradigms offers the potential to improve cognition, mood and daily functioning, but its effectiveness is not well established. [12] The objective of this research was to conduct a systematic review and meta-analysis to determine the efficacy of CCT in depressive disorders.

Other applications like a cognitive studies is to give that working memory is an important cognitive skill that is linked to academic success, there is increasing attention to exploring forms to support working memory problems in students [13]. A promising approach is computerized training, and the objective of the present study focused on a computerized training of working memory that could lead to effects in the training process. Students were assigned to one of three groups: Trained a once a week (WMT-low frequency); Study Group, where they trained four times a week (high-WMT frequency). All three groups were tested on memory measures, verbal and nonverbal ability, and academic achievement of work before training; and re-tested in the same measures after training, as well as 8 months later. The data indicate increases in both verbal and visuo-spatial working memory tasks for the high frequency training group. The improvements were also evidenced in tests of verbal and nonverbal ability tests, as well as spelling, in the high frequency training group. There were some maintenance effects when the students were rehearsed 8 months later. Possible reasons as to why computerized work memory training resulted in some far transfer effects in the high frequency training group are included in the debate.

The computerized auditory cognitive training to improve cognition and functional outcomes in patients with heart failure was a pilot study, focused on the feasibility and effectiveness of auditory computerized cognitive training (ACT) was examined in patients with heart failure (HF) [14]. Individuals with HF have four times the risk of cognitive impairment, but cognitive intervention studies are scarce, making use of computerized intervention to work attention and memory issues.

For the specific case of this study, the cyclon application is analyzed in the light of a parallel experiment with the same population of influence with another application called GlyphReader, which aims at training in phonemes for children with cognitive and intellectual disabilities.

4   Methodology

The experience of implementing the Ondas program shows a diversity of approaches and methodological processes impossible to synthesize in a single methodological commit-ment, since each group mixed tools and components of different approaches, depending on the problems worked, the advisors, the possibilities of resources and access to certain types of instruments, according to the context in which it will be developed [15].

The foregoing allowed us to know that the research response is experienced within a perspective of "epistemological and methodological freedom of investigation. The teams define their epistemological perspective - empirical-analytical, hermeneutic, social critic, constructivist, among others - as well as its methodological approach - quantitative, qualitative, participatory, etc. - based on its relevance to answer the question and the objectives of the research. It also recognizes the differences in the research styles of the subjects and in their structures of thought and training [16, 17].

The specialized accompaniment, developed by the external advisors of the projects and by the teachers, has made possible the process of systematic inquiry oriented from different approaches and has translated these elaborate and complex systems into the logic of children's and youth cultures, without losing the conceptual and procedu-ral rigor; element that has allowed the emergence of this new field of knowledge: to build research processes in initial education for the younger age groups of the population [18, 19].

It has also been concluded that one of the aspects that permanently makes a presence in the different methodological approaches used, is the question as a starting point and feedback of the research process. This question also has some special characteristics, and it can be affirmed that in the 6 years of development its implementation there are some accumulated on it: its use, its methodological place and its conception, which today allow it to be recognized as central or base of the methodological strategy.

Consequently, the methodological process and the place of the question "are understood not as instrumental processes that operate mechanically, but with culturally situated subjects, who put into play their sensitivity, their knowledge, their prejudices, their ability to observe, of creating and innovating in the research process, specifically of these age groups" [20].

Therefore, it is intended that the virtual community knows different research methods and their way of applying them, that integrates methods and techniques to collect information so as to overcome old contradictions and fragmentations of reality between the natural and the social, between the narrative and the quantitative, and that, in collective work, agreements are generated that allow defining their research paths, consistent with their projects and in relation to the specific needs of the regions, helping to build the specificity of Waves [21, 22].

4.1 Population and Sampling

For the development of this research, 350 first-grade students were taken from the public schools of the Magdalena department in Colombia, who were intervened in the reading-writing processes through the development of this project.

5 Applications Features

Cyclon first grade is a software tool that supports the identification and learning of the vowels and consonants of the first grade children of the educational institutions of the Magdalena department in Colombia. Through this implementation it is sought to make a learning of the Spanish mother tongue as a fundamental basis for the use of the reading-writing process in students, see Fig. 1.

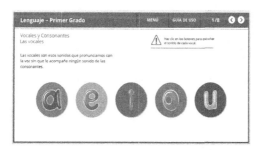

Fig. 1. Learning the vowels

Fig. 2. Listening vowels

Additionally, they are included through interactive activities to identify each of the letters of the alphabet to identify the correct sounds according to the different activities proposed, see Fig. 2.

Additionally, each of the users can select both vowels and consonants to form different words at the beginning of the reading process (Fig. 3).

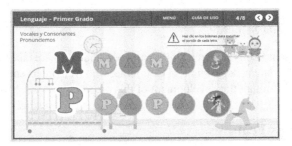

Fig. 3. Finding new words

6 Results and Conclusions

As a result of the software implementation, it was found that the use of information technologies and communications contribute to the learning process, especially in children who begin the schooling process. The results before the intervention of the software and after the use of the application can be seen in the following table and are compared with the use of the GlyphReader tool (Tables 1 and 2).

Table 1. Variable analysis with cyclon language app

Variable	Results before using the tool	Results after using the tool
Vowel identification	50%	80%
Reading sounds	40%	85%
Find new words	55%	95%

Table 2. Variable analysis with GlyphReader

Variable	Results before using the tool	Results after using the tool
Vowel Identification	50%	70%
Reading sounds	40%	75%
Find new words	55%	85%

Performing the analysis of the sensitive variables of the study, it can be noted that in identification of vowels taking into account the methodology and definition of activities, Cyclon App obtains an improvement of 30% versus GlyphReader with 20%, which can be observed in the Fig. 4.

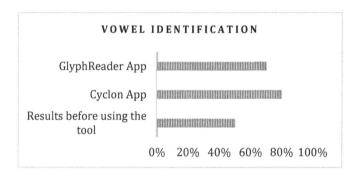

Fig. 4. Variable comparison vowel identification

As far as reading of sounds could be identified, Cyclon App obtains an improvement of 45% versus GlyphReader with 35%, which can be seen in Fig. 5.

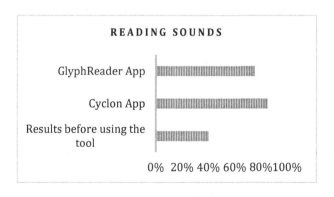

Fig. 5. Variable comparison reading sounds

In the competition to find new words, Cyclon App obtains an improvement of 45% versus GlyphReader with 35%, which can be seen in Fig. 6.

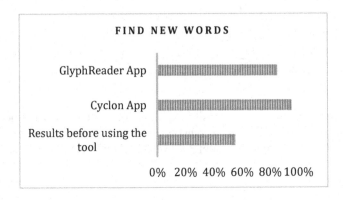

Fig. 6. Variable comparison finding new words

References

1. Laukkanen, A., Pesola, A.J., Heikkinen, R., Sääkslahti, A.K., Finni, T.: Family-based cluster randomized controlled trial enhancing physical activity and motor competence in 4–7-year-old children. PLoS ONE **10**(10), e0141124 (2015)
2. Domitrovich, C.E., Durlak, J.A., Staley, K.C., Weissberg, R.P.: Social-emotional competence: an essential factor for promoting positive adjustment and reducing risk in school children. Child Dev. **88**(2), 408–416 (2017)
3. Hartmann, K., et al.: Cutaneous manifestations in patients with mastocytosis: consensus report of the European competence network on mastocytosis; the American academy of allergy, asthma & immunology; and the European academy of allergology and clinical immunology. J. Allergy Clin. Immunol. **137**(1), 35–45 (2016)
4. Nufiar, N., Idris, S.: Teacher competence test of Islamic primary teachers education in state Islamic primary schools (MIN) of Pidie regency. Jurnal Ilmiah Peuradeun **4**(3), 309–320 (2016)
5. Hanko, G.: Increasing competence through collaborative problem-solving: using insight into social and emotional factors in children's learning. David Fulton Publishers, London (2016)
6. Laukkanen, A.: Physical Activity and Motor Competence in 4–8-Year Old Children: Results of a Family-Based Cluster-Randomized Controlled Physical Activity Trial. Studies in Sport, Physical Education and Health 238. University of Jyväskylä, Jyväskylä (2016)
7. de Araújo Vilhena, D., Sucena, A., Castro, S.L., Pinheiro, Â.M.V.: Reading test—sentence comprehension: an adapted version of Lobrot's lecture 3 test for Brazilian Portuguese. Dyslexia **22**(1), 47–63 (2016)
8. Peters, R., Broekens, J., Neerincx, M.A.: Robots educate in style: the effect of context and non-verbal behaviour on children's perceptions of warmth and competence. In: 2017 26th IEEE International Symposium on Robot and Human Interactive Communication (RO-MAN), pp. 449–455). IEEE, August 2017

9. Ke, Z., Borakova, N.U., Valiullina, G.V.: Peculiarities of psychological competence formation of university teachers in inclusive educational environment. Eurasia J. Math. Sci. Technol. Educ. **13**(8), 4701–4713 (2017)

10. Rabipour, S., Davidson, P.S.: Do you believe in brain training? A questionnaire about expectations of computerised cognitive training. Behav. Brain Res. **295**, 64–70 (2015)

11. Pressler, S.J., et al.: Nurse-enhanced computerized cognitive training increases serum brain-derived neurotropic factor levels and improves working memory in heart failure. J. Cardiac Fail. **21**(8), 630–641 (2015)

12. Motter, J.N., Pimontel, M.A., Rindskopf, D., Devanand, D.P., Doraiswamy, P.M., Sneed, J.R.: Computerized cognitive training and functional recovery in major depressive disorder: a meta-analysis. J. Affect. Disord. **189**, 184–191 (2016)

13. Alloway, T.P., Bibile, V., Lau, G.: Computerized working memory training: can it lead to gains in cognitive skills in students? Comput. Hum. Behav. **29**(3), 632–638 (2013)

14. Athilingam, P., Edwards, J.D., Valdes, E.G., Ji, M., Guglin, M.: Computerized auditory cognitive training to improve cognition and functional outcomes in patients with heart failure: results of a pilot study. Heart Lung **44**(2), 120–128 (2015)

15. Wickman, K., Nordlund, M., Holm, C.: The relationship between physical activity and self-efficacy in children with disabilities. Sport Soc. **21**(1), 50–63 (2018)

16. Snow, P.C.: Elizabeth Usher Memorial Lecture: language is literacy is language-positioning speech-language pathology in education policy, practice, paradigms and polemics. Int. J. Speech Lang. Pathol. **18**(3), 216–228 (2016)

17. Shek, D.T., Yu, L., Siu, A.M.: Interpersonal competence and service leadership. Int. J. Disabil. Hum. Dev. **14**(3), 265–274 (2015)

18. Arbour, M., Kaspar, R.W., Teall, A.M.: Strategies to promote cultural competence in distance education. J. Transcult. Nurs. **26**(4), 436–440 (2015)

19. Sandler, J.: Dimensions of Psychoanalysis: A Selection of Papers Presented at the Freud Memorial Lectures. Routledge, Abingdon (2018)

20. De-La-Hoz-Franco, E., Ariza-Colpas, P., Quero, J.M., Espinilla, M.: Sensor-based datasets for human activity recognition–a systematic review of literature. IEEE Access **6**, 59192–59210 (2018)

21. Ariza, P., Pineres, M., Santiago, L., Mercado, N., De la Hoz, A.: Implementation of moprosoft level I and II in software development companies in the colombian caribbean, a commitment to the software product quality region. In: 2014 IEEE Central America and Panama Convention (CONCAPAN XXXIV), pp. 1–5. IEEE, November 2014

22. Ariza-Colpas, P., Oviedo-Carrascal, A.I., De-la-hoz-Franco, E.: Using k-means algorithm for description analysis of text in RSS news format. In: Tan, Y., Shi, Y. (eds.) DMBD 2019. CCIS, vol. 1071, pp. 162–169. Springer, Singapore (2019). https://doi.org/10.1007/978-981-32-9563-6_17

23. Piñeres-Melo, M.A., Ariza-Colpas, P.P., Nieto-Bernal, W., Morales-Ortega, R.: SSwWS: structural model of information architecture. In: Tan, Y., Shi, Y., Niu, B. (eds.) ICSI 2019. LNCS, vol. 11656, pp. 400–410. Springer, Cham (2019). https://doi.org/10.1007/978-3-030-26354-6_40

24. Ariza-Colpas, P.P., Piñeres-Melo, M.A., Nieto-Bernal, W., Morales-Ortega, R.: WSIA: web ontological search engine based on smart agents applied to scientific articles. In: Tan, Y., Shi, Y., Niu, B. (eds.) ICSI 2019. LNCS, vol. 11656, pp. 338–347. Springer, Cham (2019). https://doi.org/10.1007/978-3-030-26354-6_34

25. Ariza-Colpas, P., et al.: Enkephalon - technological platform to support the diagnosis of alzheimer's disease through the analysis of resonance images using data mining techniques. In: Tan, Y., Shi, Y., Niu, B. (eds.) ICSI 2019. LNCS, vol. 11656, pp. 211–220. Springer, Cham (2019). https://doi.org/10.1007/978-3-030-26354-6_21

26. Ariza-Colpas, P., Morales-Ortega, R., Piñeres-Melo, M., De la Hoz-Franco, E., Echeverri-Ocampo, I., Salas-Navarro, K.: Parkinson disease analysis using supervised and unsupervised techniques. In: Tan, Y., Shi, Y., Niu, B. (eds.) ICSI 2019. LNCS, vol. 11656, pp. 191–199. Springer, Cham (2019). https://doi.org/10.1007/978-3-030-26354-6_19

Quadrotor Modeling and a PID Control Approach

César A. Cárdenas R.[1(✉)], Víctor Hugo Grisales[2],
Carlos Andrés Collazos Morales[3], H. D. Cerón-Muñoz[4], Paola Ariza-Colpas[5],
and Roger Caputo-Llanos[5]

[1] Universidad Manuela Beltrán, Bogotá D.C., Colombia
`cesar.cardenas@docentes.umb.edu.co`
[2] Departamento de Ingeniería Mecánica y Mecatrónica,
Universidad Nacional de Colombia, Bogotá D.C., Colombia
[3] Grupo de Ciencias Básicas y Laboratorios, Universidad Manuela Beltrán,
Bogotá D.C., Colombia
`carlos.collazos@docentes.umb.edu.co`
[4] Department of Aeronautical Engineering, University of São Paulo,
São Paulo, Brazil
[5] Departamento de Ciencias de la Computación y Electrónica,
Universidad de la Costa, Barranquilla, Colombia

Abstract. Since there has been an important increase in unmanned
vehicles systems research such as quadrotors, a mathematical model and
PID control laws are studied. Based on some dynamic variables, PID
control is applied to compute a controller to be then use in autopilot
simulations. As this kind of VTOL vehicle seems to be unstable, the aim
of this work is to change even other flight mechanics parameters and
control gains to study attitude and altitude variations. A well-known
computational tool is used for simulation purposes, performance analysis
and validation.

Keywords: Quadrotor · PID control · VTOL · Flight dynamics

1 Introduction

Unmanned aerial devices have been developing significant capacities to fly.
Therefore, many researchers from multidisciplinaty areas have shown interest
in aerial vehicles in which the human influence can be greatly reduced. Different engineering areas such as aerodynamics, control, path planning and for
instance autonomous platforms play an important role in this research area.
These type of flying vehicles can be designed nowadays to carry or hold different
payloads. Likewise, path planning and path following control are being utilized

The original version of this chapter was revised: An incorrect version of an author's
affiliation was published. This has been corrected. The correction to this chapter is
available at https://doi.org/10.1007/978-3-030-44689-5_26

to achieve autonomous tasks. A multirotor type that has vertical takeoff and landing (VTOL) capabilities is the quadrotor. Currently, these kind of vehicles have a large use in aerospace businesses. They can fly autonomously to perform several missions such as environmental research, rescue, traffic or infrastructure inspections, agricultural monitoring, image and video, scientific research, inspections of places with very difficult access and even products delivery. Hence, their uses are not just restricted to dangerous works. [9,10].

2   Reference Frames

For a math description model and control issues, at least one reference frame is required. If other reference frames are employed, it will be easier to get the derivation of the motion equations. Once multiple reference frames are used, important matters are related to vector transformations from one frame to another. The rotation matrices are based on Euler angles. A reference fixed frame is taken to obtain distance and direction. A coordinate system is considered for measurements. There are two reference frames clearly defined, the Earth Fixed Frame and Body Fixed Frame. [3,6,7,10] The E-frame is selected as the inertial frame. (0_E X_E Y_E Z_E). Its origin is at O_E. In this reference frame, both the linear position (ξ^E) and angular position (Θ^E) are defined. There is another reference frame required as the body frame (0_B X_B Y_B Z_B) and is attached to the body of the quadrotor. The origin is at the reference point O_B. In this B-frame some dynamic variables such as the linear velocity, the angular velocity and the forces and torques are determined. (Fig. 1).

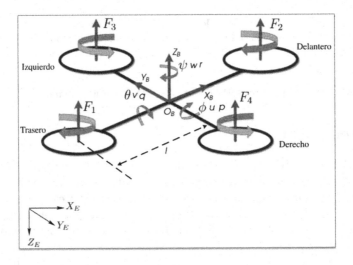

Fig. 1. Reference systems

3   Quadrotor System Variables

A quadrotor commonly has six degrees of freedom (6DOF) and are defined as: $\xi = (x, y, z)$ represents the linear position and $\Theta = (\theta, \phi, \psi)$ is the attitude or also known as Euler angles *pitch, roll* and *yaw*. Thus, if $\Theta = (\theta, \phi, \psi)$ and $\xi = (x, y, z)$, we obtain the general position vector Φ as:

$$\Phi = \begin{bmatrix} \xi \\ \Theta \end{bmatrix} = \begin{bmatrix} x \\ y \\ z \\ \theta \\ \phi \\ \psi \end{bmatrix} \tag{1}$$

The following are the linear and angular velocities $v = (u, v, w)\ \omega = (p, q, r)$ are:

$$\upsilon = \begin{bmatrix} v \\ \omega \end{bmatrix} = \begin{bmatrix} u \\ v \\ w \\ p \\ q \\ r \end{bmatrix} \tag{2}$$

3.1   Mathematical Model

For vector transformations from the E-frame to the B-frame, there is a matrix which is known as the direction cosine matrix. If the rotations are first around x axis next about y and the final one around z axis, the rotation matrix got is: $R_\Upsilon = R(\phi, \theta, \psi) = R_x(\phi)R_y(\theta)R_z(\psi)$. [7,8,10,11]

$$R_x(\phi) = \begin{bmatrix} 1 & 0 & 0 \\ 0 & cos\phi & -sin\phi \\ 0 & sin\phi & cos\phi \end{bmatrix} \tag{3}$$

$$R_y(\theta) = \begin{bmatrix} cos\theta & 0 & sin\theta \\ 0 & 1 & 0 \\ -sin\theta & 0 & cos\theta \end{bmatrix} \tag{4}$$

$$R_z(\psi) = \begin{bmatrix} cos\psi & -sin\psi & 0 \\ sin\psi & cos\psi & 0 \\ 0 & 0 & 1 \end{bmatrix} \tag{5}$$

Hence, the rotation matrix product is as follows:

$$R_{\Upsilon_B^E} = \begin{bmatrix} cos\theta cos\psi & cos\psi sin\theta sin\phi - sin\psi cos\phi & sin\phi sin\psi + cos\phi cos\psi sin\theta \\ sin\psi cos\theta & cos\psi cos\phi + sin\psi sin\theta sin\phi & sin\psi sin\theta cos\phi - cos\psi sin\phi \\ -sin\theta & cos\theta sin\phi & cos\phi cos\theta \end{bmatrix} \quad (6)$$

If smalls angles of motion are taken $[\dot{\phi}\ \dot{\theta}\ \dot{\psi}]^T = [p\ q\ r]^T$, therefore, the dynamic model is: [7,8]

$$\begin{cases} \ddot{x} = \frac{F_{total}}{m}[sin(\phi)sin(\psi) + cos(\phi)cos(\psi)sin(\theta)] \\ \ddot{y} = \frac{F_{total}}{m}[cos(\phi)sin(\psi)sin(\theta) - cos(\psi)sin(\phi)] \\ \ddot{z} = -g\frac{F_{total}}{m}[cos(\phi)cos(\theta)] \\ \ddot{\phi} = \frac{I_{yy}-I_{zz}}{I_{xx}}\dot{\theta}\dot{\psi} + \frac{\tau_x}{I_{xx}} \\ \ddot{\theta} = \frac{I_{zz}-I_{xx}}{I_{yy}}\dot{\phi}\dot{\psi} + \frac{\tau_x}{I_{yy}} \\ \ddot{\psi} = \frac{I_{xx}-I_{yy}}{I_{zz}}\dot{\phi}\dot{\theta} + \frac{\tau_x}{I_{zz}} \end{cases} \quad (7)$$

4   Control Model of a Quadrotor

Four torques are needed to control a quadrotor ($\tau_1\ \tau_2\ \tau_3\ \tau_4$), which are given to the motors to generate the thrust forces ($F_1\ F_2\ F_3\ F_4$) in z. The net torques acting on the body can be calculated by using the inputs $U = (U_1\ U_2\ U_3\ U_4)$ that can be applied to control the quadrotor. As an aerodynamic consideration forces and torques are proportional to the squared propeller's speed. [1]. Therefore the relationship between motions and propellers' squared speed is as follows:

$$\begin{aligned} F_T &= b(\Omega_1^2 + \Omega_2^2 + \Omega_3^2 - \Omega_4^2) \\ \tau_{x(\theta)} &= bl(\Omega_2^2 - \Omega_4^2) \\ \tau_{y(\phi)} &= bl(\Omega_1^2 - \Omega_3^2) \\ \tau_{z(\psi)} &= K_{drag}(\Omega_1^2 + \Omega_2^2 - \Omega_3^2 + \Omega_4^2) \end{aligned} \quad (8)$$

The motor $\Omega\ [rad\ s^{-1}]$ and vector speeds $\boldsymbol{\Omega}$ are the following: [1].

$$\boldsymbol{\Omega} = \begin{bmatrix} \Omega_1 \\ \Omega_2 \\ \Omega_3 \\ \Omega_4 \end{bmatrix} \quad (9)$$

$\Omega_1\ [rad\ s^{-1}]$ and $\Omega_3\ [rad\ s^{-1}]$ are the motor speeds (front and rear), also Ω_2 $[rad\ s^{-1}]$ and $\Omega_4\ [rad\ s^{-1}]$ are the right and left motor speeds. The relationship between the control inputs and the speeds of the motor is:

$$\begin{bmatrix} U_1 \\ U_2 \\ U_3 \\ U_4 \end{bmatrix} = \begin{bmatrix} b(\Omega_1^2 + \Omega_2^2 + \Omega_3^2 + \Omega_4^2) \\ bl(\Omega_4^2 - \Omega_2^2) \\ bl(\Omega_3^2 - \Omega_1^2) \\ d(\Omega_2^2 + \Omega_4^2 - \Omega_1^2 - \Omega_3^2) \end{bmatrix} \quad (10)$$

l [m] is the quadrotor center distance and $[U_1\ U_2\ U_3\ U_4]$ are the control inputs. U_1 is the thrust force responsible for altitude z. U_2 is the thrust difference between propellers 2 and 4 that originates the roll moment. U_3 represents the thrust variation between motors 1 and 3 that generate the pitch moment. Finally, U_4 is the mixing of the single torques between the clockwise and counterclockwise rotors that are responsible for yaw rotation. Similarly, U_1 creates the desired altitude. U_2 and U_3 generate the respective roll and pitch angles and U_4 originates the yaw angle. [1,4,5]. In addition, the linearization about a point of equilibrium is: $\dot{\mathbf{x}} = f(x, u)$. Hence, as a linear model we obtain: [1,5,11].

$$\mathbf{f(x,u)} = \begin{cases} \dot{\phi} = p \\ \dot{\theta} = q \\ \dot{\psi} = r \\ \dot{p} = \frac{\tau_x + \tau_{ax}}{I_x} \\ \dot{q} = \frac{\tau_y + \tau_{ay}}{I_y} \\ \dot{r} = \frac{\tau_z + \tau_{az}}{I_z} \\ \dot{u} = -g\theta + \frac{F_{ax}}{m} \\ \dot{v} = g\phi + \frac{F_{ay}}{m} \\ \dot{w} = \frac{F_{az} - F_{total}}{m} \\ \dot{x} = u \\ \dot{y} = v \\ \dot{z} = w \end{cases} \qquad (11)$$

5   Simplified Control Model

A quadrotor is certainly an underactuated system. It is a 6DOF machine with just four control inputs. So, it may be really difficult to control 6 degrees of freedom having only four control inputs. Mainly, the control work is carried out on attitude and the yaw angle. Hence, the reduced dynamics model considered for control is:

$$\begin{pmatrix} \ddot{z} = g - (cos\phi cos\theta)\frac{1}{m}U_1 \\ \ddot{\phi} = \frac{U_2}{I_x} \\ \ddot{\theta} = \frac{U_3}{I_y} \\ \ddot{\psi} = \frac{U_4}{I_z} \end{pmatrix} \qquad (12)$$

5.1   Control Laws for Attitude and Altitude

The following PID control law is the taken for altitude:

$$U_1 = K_p^z e_z + K_d^z \dot{e}_z + K_i^z \int (e_z), \ e_z = z - z_d \qquad (13)$$

K_p, K_d, K_i are the gains (proportional, derivative and integral), $\dot{e}_z$ represents the desired altitude change and z_d is defined as the desired altitude. If matrix 10 is inverted, we get the relationship between U and Ω^2 as follows:

$$
\begin{bmatrix} \Omega_1^2 \\ \Omega_2^2 \\ \Omega_3^2 \\ \Omega_4^2 \end{bmatrix} = \begin{bmatrix} \frac{1}{4b} & 0 & \frac{1}{2bl} & -\frac{1}{4d} \\ \frac{1}{4b} & -\frac{1}{2bl} & 0 & \frac{1}{4d} \\ \frac{1}{4b} & 0 & \frac{1}{2bl} & -\frac{1}{4d} \\ \frac{1}{4b} & \frac{1}{2bl} & 0 & \frac{1}{4d} \end{bmatrix} \begin{bmatrix} U_1 \\ U_2 \\ U_3 \\ U_4 \end{bmatrix}
\tag{14}
$$

Also, the control laws for attitude or orientation angles (roll, pitch and yaw) are taken as:

$$
U_2 = K_p^\phi e_\phi + K_d^\phi \dot{e}_\phi + K_i^\phi \int (e_\phi),\ e_\phi = \phi - \phi_d
\tag{15}
$$

$$
U_3 = K_p^\theta e_\theta + K_d^\theta \dot{e}_\theta + K_i^\theta \int (e_\theta),\ e_\theta = \theta - \theta_d
\tag{16}
$$

$$
U_4 = K_p^\psi e_\psi + K_d^\psi \dot{e}_\psi + K_i^\psi \int (e_\psi),\ e_\psi = \psi - \psi_d
\tag{17}
$$

Similarly, K_p, K_d, K_i are the proportional, derivative and integral gains, $\dot{e}_\phi, \dot{e}_\theta, \dot{e}_\psi$ are the adjustments for the desired angles and ϕ_d, θ_d, ψ_d are the desired angles.

5.2   Transfer Functions

The following are the attitude transfer functions:

$$
G_{\Phi/U_2}(s) = \frac{1}{s^2 I_x}
$$

$$
G_{\Theta/U_3}(s) = \frac{1}{s^2 I_y}
\tag{18}
$$

$$
G_{\Psi/U_4}(s) = \frac{1}{s^2 I_z}
$$

And for altitude we have,

$$
G_{z/U_1}(s) = \frac{1}{s^2\, m}
\tag{19}
$$

6   PID Controller Computation

The transfer function considered for a PID controller is (Table 1):

$$
G(s) = P + \frac{I}{s} + \frac{Ds}{s+1}
\tag{20}
$$

Some autopilot devices just implement PD control. Thus, the transfer function is as follows:

$$G(s) = P + \frac{Ds}{s+1} \tag{21}$$

Table 1. Controller design parameters

PD-PID parameters
$\zeta = 0.6 \quad ts = 5s \quad \beta = 10\zeta\omega_n$

The PD controller values are:

$$P = \omega_n^2 I_{xyz}$$
$$D = 2\zeta\omega_n I_{xyz}$$

For the PID controllers are given as:

$$P = 2\zeta\omega_n\beta + \omega_n^2 I_{xyz}$$
$$I = \omega_n^2\beta I_{xyz}$$
$$D = 2\zeta\omega_n\beta I_{xyz}$$

Thus, the controller variables are:

$$P = m\omega_n^2$$
$$D = 2\zeta\omega_n m$$

Finally, the parameters for the PID controller are given by (Table 2):

$$P = 2\zeta\omega_n\beta + \omega_n^2 m$$
$$I = \omega_n^2\beta m$$
$$D = 2\zeta\omega_n + \beta m$$

Other variables are also taken for control simulation (Table 3) and (Figs. 2, 3, 4, 5, 6 and 7).

Table 2. Dynamics values for simulation

Dynamic variables	Model 1	Model 2	Model 3
m	0.1 kg	0.09 kg	0.14 kg
I_x	0.45 kg · m^2	0.45 kg · m^2	0.45 kg · m^2
I_y	0.51 kg · m^2	0.51 kg · m^2	0.51 kg · m^2
I_z	0.95 kg · m^2	0.95 kg · m^2	0.95 kg · m^2
l	0.5 m	0.95 kg · m^2	0.35 m

Table 3. Dynamic parameters

Simulation control variables
$J_m = 4e^-7 \, \text{kg m}^2$
$b = 3.13e^-5 - 1.33e^-5 \, \text{N s}^2$
$d = 3.13e^-5 - d = 1.3e^-5 \, \text{N m s}^2$

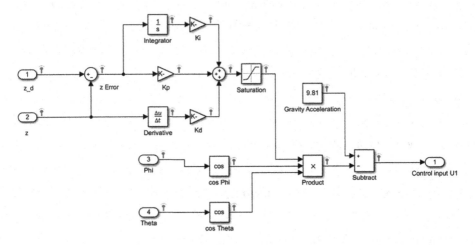

Fig. 2. Altitude PID control

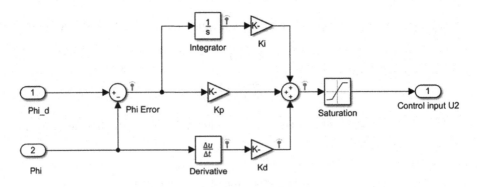

Fig. 3. Roll PID control

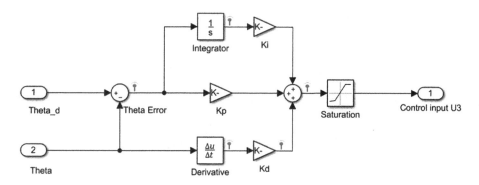

Fig. 4. Pitch PID control

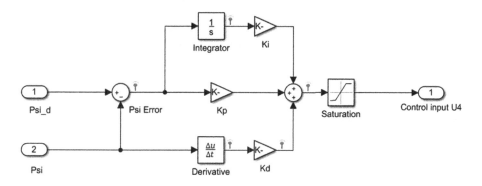

Fig. 5. Control PID yaw

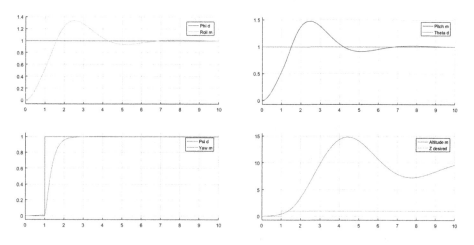

Fig. 6. Altitude/attitude PD controller response for models 1, 2, 3

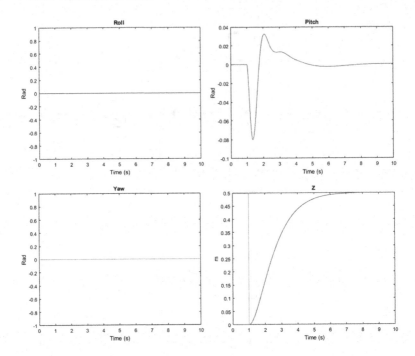

Fig. 7. Altitude/attitude PID controller response for models 1, 2, 3

7   Results

The dynamics values are separately taken for all models. It can be noticed that in the PD control simulations for models 1, 2 and 3, there is initially some overshoot. However, after some adjustments a proper attitude stabilization is achieved. Regarding to altitude performance, an important overshoot is also evident and the control signal obtained tends to deviate from the desired. This issue may appear due to some kind of altitude increase or some PID gains tuning variation still needed. For models 4, 5 and 6, a superior PD and PID control capability is noted. Some very small peaks are originated for attitude. Although, these signals are very similar or not so far from to the desired ones. A slow reaction can be noticed for the yaw angle. In spite of this fact, finally the vehicle could stabilize. The mean of the dynamics values taken is acceptable. Despite this, it is definitely necessary to make some additional changes. Some simulations show that there are some kind of changes or oscillations especially for position. It seems that there are some small oscillations when the inertia moments values are increased. Moreover, oscillations rise as the moments of inertia are small. Therefore, this the fact could be created because of some abrupt changes in attitude. These changes are perceived in simulation. It is possible to be associated to the increase or reduction of the thrust force and configuration type $(X, +)$.

8   Conclusions

A mathematical model of a quadrotor is provided based on the Newton-Euler method. Another mathematical and linearization approach for further studies may be considered with Taylor series as presented in [2,10]. Besides the analytical computation of the PID gains, it is required to use a computational tool such as MATLAB (PID option) to adjust or get more precise gains values for better controller performance. It seems that it is feasible to accomplish appropriate results with just a PD controller. It might be suitable for a stable flight (attitude). Although some very small peaks are observed in attitude control signals, they are not far from the desired settings. It is also seen a slow response of the yaw angle. However, after a while the platform can be stabilized. The altitude behaviour is not so right and the state error increases. On the other hand, a much better altitude and attitude performance is got for models 4, 5, and 6. Some negative estimations are acquired for orientation angles. This may be possible due to motors configuration $(X, +)$. These values just give an inclination and don't affect the angle magnitude. It is inferred that the dynamic parameters certainly can have an significant effect on the PID controller efficiency.

References

1. Bresciani, T.: Modelling, identification and control of a quadrotor helicopter. M.Sc. theses (2008)
2. Collazos, C., et al.: State estimation of a dehydration process by interval analysis. In: Figueroa-García, J.C., López-Santana, E.R., Rodriguez-Molano, J.I. (eds.) WEA 2018. CCIS, vol. 915, pp. 66–77. Springer, Cham (2018). https://doi.org/10.1007/978-3-030-00350-0_6
3. Cork, L.R.: Aircraft dynamic navigation for unmanned aerial vehicles. Ph.D. thesis, Queensland University of Technology (2014)
4. Cowling, I.: Towards autonomy of a quadrotor UAV (2008)
5. Habib, M.K., Abdelaal, W.G.A., Saad, M.S., et al.: Dynamic modeling and control of a Quadrotor using linear and nonlinear approaches (2014)
6. Lugo-Cárdenas, I., Flores, G., Salazar, S., Lozano, R.: Dubins path generation for a fixed wing UAV. In: 2014 International Conference on Unmanned Aircraft Systems (ICUAS), pp. 339–346. IEEE (2014)
7. Mulder, S.: Flight Dynamics. Princeton University Press, Princeton (2007)
8. Swartling, J.O.: Circumnavigation with a group of quadrotor helicopters (2014)
9. Pharpatara, P.: Trajectory planning for aerial vehicles with constraints. Theses, Université Paris-Saclay, Université d'Evry-Val-d'Essonne, September 2015. https://tel.archives-ouvertes.fr/tel-01206423
10. Poyi, G.T.: A novel approach to the control of quad-rotor helicopters using fuzzy-neural networks (2014)
11. Sabatino, F.: Quadrotor control: modeling, nonlinearcontrol design, and simulation (2015)

Correction to: Quadrotor Modeling and a PID Control Approach

César A. Cárdenas R., Víctor Hugo Grisales, Carlos Andrés Collazos
Morales, ·H. D. Cerón-Muñoz, Paola Ariza-Colpas,
and Roger Caputo-Llanos

Correction to:
Chapter "Quadrotor Modeling and a PID Control Approach"
in: U. S. Tiwary and S. Chaudhury (Eds.): *Intelligent Human*
Computer Interaction, **LNCS 11886,**
https://doi.org/10.1007/978-3-030-44689-5_25

The chapter was inadvertently published with an incorrect version of an author's affiliation as "Universidad Manuel Beltrán" whereas it should correctly read "Universidad Manuela Beltrán".

The updated version of this chapter can be found at
https://doi.org/10.1007/978-3-030-44689-5_25

© Springer Nature Switzerland AG 2020
U. S. Tiwary and S. Chaudhury (Eds.): IHCI 2019, LNCS 11886, p. C1, 2020.
https://doi.org/10.1007/978-3-030-44689-5_26

Author Index